Grade 7 · Unit 3

Inspire
Science

Earth's Resources

Mc
Graw
Hill
Education

Phenomenon: What are these pools used for?

This image shows an area where lithium is mined. Lithium is an element and is the 33rd most abundant element on Earth.

Although lithium is a metal, it is soft enough to cut with a butter knife!

FRONT COVER: (t)alekosa/iStock/Getty Images, (b)Abraksis/iStock/Getty Images.
BACK COVER: alekosa/iStock/Getty Images.

mheducation.com/prek-12

Send all inquiries to:
McGraw-Hill Education
STEM Learning Solutions Center
8787 Orion Place
Columbus, OH 43240

ISBN: 978-0-07-687457-6
MHID: 0-07-687457-5

Printed in the United States of America.

3 4 5 6 7 QVS 22 21 20 19

McGraw-Hill is committed to providing instructional materials in Science, Technology, Engineering, and Mathematics (STEM) that give all students a solid foundation, one that prepares them for college and careers in the 21st century.

Welcome to

Inspire Science

Explore Our Phenomenal World

Learning begins with curiosity. Inspire Science is designed to spark your interest and empower you to ask more questions, think more critically, and generate innovative ideas.

Start exploring now!

Inspire Curiosity • Inspire Investigation • Inspire Innovation

Authors, Contributors, and Partners

Program Authors

Alton L. Biggs
Biggs Educational Consulting
Commerce, TX

Ralph M. Feather, Jr., PhD
Professor of Educational Studies and
Secondary Education
Bloomsburg University
Bloomsburg, PA

Douglas Fisher, PhD
Professor of Teacher Education
San Diego State University
San Diego, CA

Page Keeley, MEd
Author, Consultant, Inventor of
Page Keeley Science Probes
Maine Mathematics and Science
Alliance
Augusta, ME

Michael Manga, PhD
Professor
University of California, Berkeley
Berkeley, CA

Edward P. Ortleb
Science/Safety Consultant
St. Louis, MO

Dinah Zike, MEd
Author, Consultant, Inventor
of Foldables®
Dinah Zike Academy, Dinah-Might
Adventures, LP
San Antonio, TX

Advisors

Phil Lafontaine
NGSS Education Consultant
Folsom, CA

Donna Markey
NBCT, Vista Unified School District
Vista, CA

Julie Olson
NGSS Consultant
Mitchell Senior High/Second Chance
High School
Mitchell, SD

Content Consultants

Chris Anderson
STEM Coach and Engineering
Consultant
Cinnaminson, NJ

Emily Miller
EL Consultant
Madison, WI

Key Partners

American Museum of Natural History

The American Museum of Natural History is one of the world's preeminent scientific and cultural institutions. Founded in 1869, the Museum has advanced its global mission to discover, interpret, and disseminate information about human cultures, the natural world, and the universe through a wide-ranging program of scientific research, education, and exhibition.

SpongeLab Interactives

SpongeLab Interactives is a learning technology company that inspires learning and engagement by creating gamified environments that encourage students to interact with digital learning experiences. Students participate in inquiry activities and problem-solving to explore a variety of topics through the use of games, interactives, and video while teachers take advantage of formative, summative, or performance-based assessment information that is gathered through the learning management system.

PhET Interactive Simulations

The PhET Interactive Simulations project at the University of Colorado Boulder provides teachers and students with interactive science and math simulations. Based on extensive education research, PhET simulations engage students through an intuitive, game-like environment where students learn through exploration and discovery.

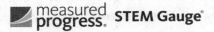

Measured Progress, a not-for-profit organization, is a pioneer in authentic, standards-based assessments. Included with New York Inspire Science is **Measured Progress STEM Gauge®** assessment content which enables teacher to monitor progress toward learning NGSS.

Table of Contents
Earth's Resources

Module 2 Materials Science

Distribution of Earth's Resources

ENCOUNTER
THE PHENOMENON

Why are salt deposits only found in certain areas on Earth?

Salt of the Earth

GO ONLINE
Watch the video *Salt of the Earth* to see this phenomenon in action.

Collaborate Salt, like other minerals, is only mined, or farmed, from certain regions worldwide. With a partner, brainstorm the geologic processes that you think are responsible for the distribution of salt and other natural resources. Why do you think Earth's resources are unevenly distributed? Record your thoughts in the space below.

Where in the world...?

Lesson 1
Natural
Resources

Lesson 2
Distribution of
Resources

Lesson 3
Depletion of
Resources

You are watching a local news report about how an unusually cold winter has brought freezing temperatures and icy conditions to several western states. The report states that rock salt, a common remedy for slippery roads, is not readily available in these states and is only found in certain parts of the world. Why do some resources only exist in certain areas?

You decide to conduct an investigation to find out 1) how people depend on Earth's resources, 2) why some regions on Earth are rich in certain resources while others are not, and 3) how people affect resource distributions. You'll use your findings to prepare a script that will be read on the 6 o'clock news explaining why resources, such as salt, are not always readily available.

Start Thinking About It

How is a scientific explanation different from a nonscientific explanation? Discuss your thoughts with your group.

STEM Module Project
Planning and Completing the Science Challenge
How will you meet this goal? The concepts you will learn throughout this module will help you plan and complete the Science Challenge. Just follow the prompts at the end of each lesson!

Renewable or Not?

Three students argued about Earth's resources. This is what they said:

Milo: Earth has an endless supply of resources.

Katia: Most of Earth's resources will eventually run out.

Greg: Some of Earth's resources can be replaced faster than they are used.

Which student do you agree with most? _____ Explain your thinking.

_____

_____

_____

_____

_____

_____

_____

You will revisit your response to the Science Probe at the end of the lesson.

Natural Resources

ENCOUNTER
THE PHENOMENON | How are humans dependent upon Earth's resources?

Within a few days, loggers can change a forest into a field of stumps. But the field may not remain barren permanently. After the stumps are cleared, new trees can be planted in their place. Trees are a valuable resource because they provide wood for fuel and construction. They also provide oxygen. Brainstorm a list of resources that come from Earth. Describe how each resource is valuable to humans. Identify which resources you think are easily replaceable or reusable over time and which ones are not.

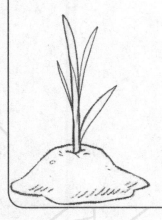

Tree-mendous
Transformation

GO ONLINE
Check out *Tree-mendous Transformation* to see this phenomenon in action.

You just brainstormed how resources, like trees, are important to humans. Make a claim about how humans depend on Earth.

CLAIM

Humans depend on Earth for...

COLLECT EVIDENCE as you work through the lesson. Then return to these pages to record your evidence.

EVIDENCE

A. What evidence have you discovered to explain how humans depend on different resources?

MORE EVIDENCE

B. What evidence have you discovered to explain the difference between renewable and nonrenewable resources?

When you are finished with the lesson, review your evidence. If necessary, based on the evidence, revise your claim.

REVISED CLAIM

Humans depend on Earth for...

Finally, explain your reasoning for how and why your evidence supports your claim.

REASONING

The evidence I collected supports my claim because...

What are natural resources?

Almost everything you use comes from natural resources. A **natural resource** is something on Earth that living things use to meet their needs. Can you identify the natural resources used to make a common object?

 Want more information?
Go online to read more about how humans depend on Earth for many different resources.

FOLDABLES

Go to the Foldables® library to make a Foldable® that will help you take notes while reading this lesson.

 Identifying Resources

Materials

classroom object

Procedure

1. Read and complete a lab safety form.

2. Choose a common object from your classroom or in your backpack.

3. In the first column of the data table below, list the object you will investigate.

4. In the second column, determine all the natural resources required to make the object. Use the Internet and other sources to gather relevant information about your object.

Data and Observations

Object	Natural Resources Required

Analyze and Conclude

5. Which natural resource was hardest to identify? How did you figure it out?

_____

_____

_____

6. Compare your data table to a classmate's. Make a class list of resources that make up objects from your classroom.

7. What type of natural resource was most common? Why might this be?

_____

_____

_____

_____

Natural Resources Life on Earth, such as this plant, requires the use of resources. The term _resource_ covers everything we use, including such basic assets as air, soil, timber, and water; fuel resources, such as coal, oil, and gas; and mineral resources, such as sand and gravel. Let's explore how humans depend on different types of natural resources.

How do humans depend on energy resources?

In the United States today, the energy used for most daily activities is easily available at the flip of a switch or the push of a button. How do you use energy in your daily activities?

 **Daily Resource Use**

Procedure

1. Design a three-column data chart in the Data and Observations section below. Title the columns *Activity, Type of Energy Used,* and *Amount of Time.*

2. Record every instance that you use energy during a 24-hr period.

3. Total your usage of the different forms of energy, and record them below.

Data and Observations

Analyze and Conclude

4. How many times did you use each type of energy?

_____

_____

_____

5. Compare and contrast your usage with that of other members of your class.

_____

_____

_____

6. Are there instances of energy use when you could have conserved energy? Explain how you would do it.

_____

_____

_____

_____

Energy Resources Energy can come from sunlight, fossil fuels, flowing water, or other sources. The power for lights, computers, appliances, and other electrical devices comes from energy resources. Nearly all manufactured products are made using energy resources. Energy resources provide the fuel you need to get from one place to another, whether it's in a car, bus, plane, or boat and can keep you warm in cold weather and cool in warm weather.

PHYSICAL SCIENCE ⟩ **Connection** Petroleum, natural gas, propane, and coal are fossil fuels. Ancient plants stored radiant energy from the Sun as chemical energy in their molecules. This chemical energy was passed on to the animals that ate the plants. Over millions of years, geological processes converted the remains of these ancient plants and animals into fossil fuels. Fossil fuels are a very concentrated form of chemical energy that easily transforms into other forms of energy.

How do humans depend on land resources?

Chances are, when you walk into your kitchen a majority of the items you touch, see, taste, and smell had roots in the land. Let's explore how items in your home are connected to land resources.

INVESTIGATION

It Comes from the Land

1. Make a list of items in a kitchen that come from land resources.

Copyright © McGraw-Hill Education Robin Matthews/Ingram Publishing

2. Design a graphic organizer to group the kitchen materials into categories.

3. Choose one category from above. How might your daily routine be altered if this category was no longer an available land resource?

_____

_____

_____

_____

_____

_____

4. Why do you think land is considered a resource?

_____

_____

_____

Land Resources People use land to grow food and for grazing animals. The wood used to make furniture, paper, cardboard, and other timber products comes from forests that cover the land. We build on land. We live and play on top of it. Land provides us with living space. People use land to create green spaces, or areas of natural vegetation in urban landscapes. Some land is also set aside for use as wilderness preserves and national parks. Development is limited in these places.

Certain minerals are mined to make products you use every day. These minerals often are called ores. **Ores** are deposits of minerals that are large enough to be mined for a profit. The average person uses 22,000 kg of mineral resources each year. For example, copper is used in electric wiring and plumbing fixtures, and quartz is used to make glass and ceramics. The automotive industry; agriculture and food production; and road, home, and building construction use mineral resources. These resources are mined from Earth.

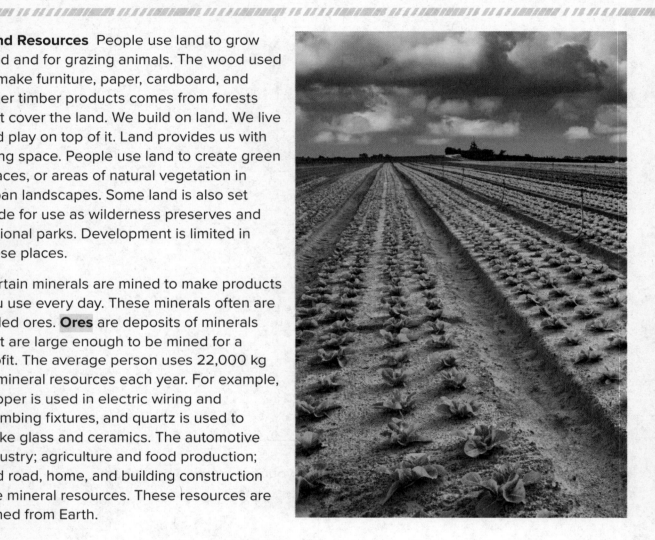

How do humans depend on air and water resources?

Water is a crucial resource for living things. Like air, we cannot survive without it. In most places in the United States, people are fortunate to have an adequate supply of clean water. When you turn on the faucet, do you think about the value of water as a resource? Let's investigate how often you use water each day.

Charted Waters

Collect data on the number of times you use water in one day. In the *Usage* column, describe how you used the water, such as *Faucet*, *Toilet*, *Shower/Bath*, *Dishwasher*, *Laundry*, *Leaks*, and *Other*. In the *Times Used* column, record and tally the total number of times you used water.

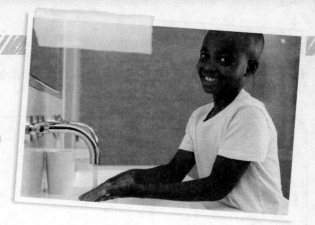

Usage	Times Used

1. **MATH Connection** Calculate the percent of water that you use for each usage category.

2. **MATH Connection** Construct a circle graph showing the percentages of each type of use in a day.

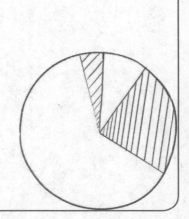

3. For which purpose did you use water the most? The least?

4. How might your household water usage change from day to day or from season to season?

_____

_____

_____

Importance of Air and Water Using some natural resources, such as fossil fuels and minerals, makes life easier. You would miss them if they were gone, but you would still survive. Air and water, on the other hand, are resources that you cannot live without. Most living things can survive only a few minutes without air. Oxygen from air helps your body provide energy for your cells.

In addition to drinking water, people use water for other purposes, including agriculture, transportation, and recreation. People use freshwater for drinking and other daily uses, such as cooking and cleaning. In addition to soil, plants need water to grow. Freshwater is used to irrigate crops. Water is also used for mining, manufacturing, and generating power. Rivers, lakes, and oceans are used to transport goods and people from place to place. Seafood is a major source of protein for many people around the world. Fish and seafood are harvested from oceans, lakes, and rivers.

LIFE SCIENCE ❭ Connection Water is necessary for the life processes of all living things, including humans. Your body is 60 percent water! Water is necessary for your cells, tissues, and organs to function. Water also helps your body regulate temperature, digest food, and perform many other functions.

COLLECT EVIDENCE

How do humans depend on different types of resources? Record your evidence (A) in the chart at the beginning of the lesson.

Desalination

Taking the Salt out of Salt Water

Anyone who's been toppled by a big ocean wave knows salt water doesn't taste like the water we drink. People can't drink salt water. It's about 200 times more salty than freshwater. About 97 percent of Earth's water is salty. Most freshwater is frozen in glaciers and ice caps, leaving less than 1 percent of the planet's water available for 7.5 billion people and countless other organisms that require freshwater to live.

The need for freshwater has scientists searching for efficient ways to take the salt out of salt water. One solution is a desalination plant, where dissolved salts are separated from seawater through a process called reverse osmosis. This is how it works:

▲ Desalination plants are found all over the world, including the United States.

❶ Salt water is pumped from the ocean.

❷ High pressure forces water through a semipermeable membrane.

❸ The semipermeable membrane acts as a filter, allowing the water, but not the salt, to pass through.

❹ Clean freshwater is collected in a separate tank.

❺ Water containing the waste salts flows out of the tank.

Copyright © McGraw-Hill Education Gregory Bull/AP Images

Because it takes a lot of energy to change salt water into freshwater, desalination plants are expensive to operate. But desalination is used in regions such as the Middle East and North Africa, where millions of people have few freshwater resources.

It's Your Turn

Journal Write a journal entry describing what you and your family can do to save water. Note which actions would be easy and which would be difficult.

AMERICAN MUSEUM
ᴼᶠ NATURAL HISTORY

Construct an explanation about how humans depend on Earth's systems for many different resources.

_____

_____

_____

How long will natural resources be around?

Every day you use energy resources when you turn on lights or play a video game. You use water resources when you brush your teeth. You eat plants grown on land. You go places in a bus or car made from minerals and powered by energy resources. And you inhale and exhale air every minute of the day. Is there an infinite supply of natural resources, or could we someday run out of them?

LAB Spill the Beans

In this activity, the red beans represent an energy resource that is available in limited amounts. The white beans represent an energy resource that is available in unlimited amounts.

Safety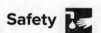

Materials

red beans (40)   paper bag

white beans (40)

Procedure

1. Read and complete a lab safety form.

2. Place 40 red beans and 40 white beans in a paper bag. Mix the contents of the bag.

Procedure, continued

3. Each team should remove 20 beans from the bag without looking at the beans. Record the numbers of red and white beans in the Data and Observations section in a data table.

4. Put the red beans aside. They are "used up." Return all the white beans to the bag. Mix the beans in the bag. Repeat steps 3 and 4 three more times.

5. Follow your teacher's directions for proper cleanup.

Data and Observations

Analyze and Conclude

6. What happened to the number of red beans drawn during each round?

_____

_____

_____

7. What would eventually happen to the red beans in the bag?

_____

_____

_____

8. How would changing the number of beans drawn in each round make the red beans last longer? Explain your answer.

_____

_____

_____

Renewable and Nonrenewable Resources Earth's resources can be classified according to how long supplies might last. **Renewable resources** are resources that can be replaced by natural processes in a relatively short amount of time. Because they are renewable, they will last a long time if used wisely. These resources come from natural processes that have been happening for billions of years and will continue to happen. Renewable resources include air, water, living things, and certain energy resources such as solar, geothermal, wind, water, and biomass.

Most energy in the United States comes from nonrenewable resources. **Nonrenewable resources** are natural resources that are being used up faster than they can be replaced by natural processes. Nonrenewable resources form slowly, usually over millions of years. If they are used faster than they form, they will run out. Nonrenewable resources include fossil fuels, such as coal, oil, natural gas, and minerals. These resources are typically limited and nonrenewable due to factors such as the long amounts of time required for them to form or the environment in which resources were created forming once or only rarely in Earth's history.

 GO ONLINE for an additional opportunity to explore!

Investigate renewable resources by performing the following activity.

☐ **Complete** a survey about the use of renewable resources in your school in the **Lab** _How are renewable energy resources used at your school?_

COLLECT EVIDENCE

What is the difference between renewable and nonrenewable resources? Record your evidence (B) in the chart at the beginning of the lesson.

Summarize It!

1. **Organize** Record information about how humans depend on each resource.

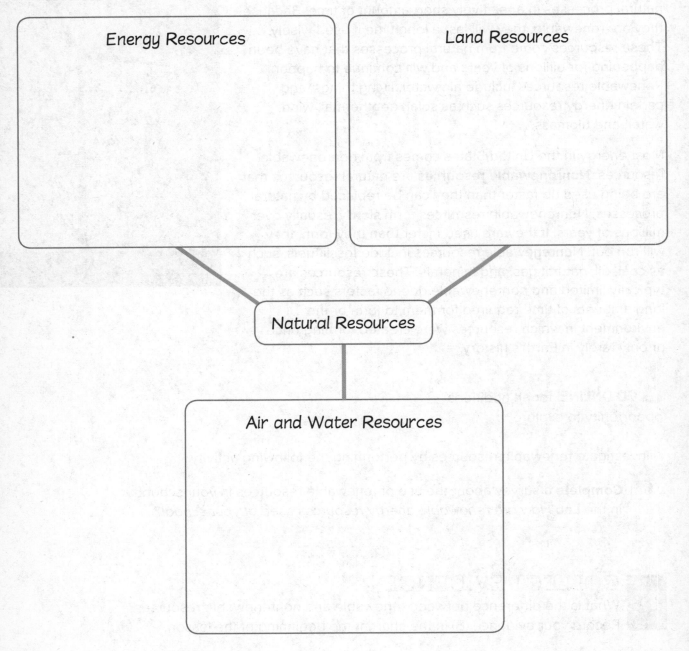

Energy Resources

Land Resources

Natural Resources

Air and Water Resources

Three-Dimensional Thinking

2. Which of the following explanations best supports why land is considered to be a resource?

   A   Using it has no consequences.

   B   It is plentiful in all parts of the world.

   C   It contains material needed by people.

   D   It produces only renewable resources.

3. When a biosphere resource is used up faster than it can be replaced, it is because

   A   it is a natural resource.

   B   it can be replaced in a short amount of time.

   C   the resource forms very quickly.

   D   the resource forms very slowly.

4. The table below shows the energy sources used to produce electricity in the United States. What can you infer from the table?

Electricity Production	
Energy Source	**Percent**
Coal	33
Natural gas	33
Solar, wind, geothermal, biomass	7
Hydroelectric power	6
Oil	1
Other	<1

   A   More electricity comes from renewable energy resources than nonrenewable resources.

   B   More electricity comes from nonrenewable energy resources than renewable resources.

   C   Hydroelectric power is more widely used for electricity than coal.

   D   Oil is more widely used for electricity than hydroelectric power.

Real-World Connection

5. Explain Why is it important to manage the use of nonrenewable resources?

_____

_____

_____

_____

6. Determine Name three items that you use every day and determine how those items are linked to resources that come from Earth.

_____

_____

_____

_____

_____

_____

 Still have questions?
Go online to check your understanding about how humans depend on Earth.

 **REVISIT SCIENCE PROBES** Do you still agree with the student you chose at the beginning of the lesson? Return to the Science Probe at the beginning of the lesson. Explain why you agree or disagree with that student now.

EXPLAIN THE PHENOMENON Revisit your claim about how humans depend upon Earth's resources. Review the evidence you collected. Explain how your evidence supports your claim.

START PLANNING STEM Module Project Science Challenge
Now that you have learned about how humans depend on Earth's resources, go to your Module Project to begin organizing information that will help you prepare your script for the news report. Keep in mind that you want to explain how people depend on Earth's resources, yet they are not always readily available.

Resource Location

Living things need natural resources to survive. These resources include mineral, energy, and groundwater resources. Where are these resources located on Earth? Circle the idea that best matches your thinking.

A: All types of natural resources can be found everywhere on Earth.

B: Natural resources are limited to certain areas on Earth.

C: Only mineral resources are limited to certain locations.

D: Mineral and energy resources are limited to certain areas, but groundwater resources are available at all locations on Earth.

Explain your thinking.

_____

_____

_____

_____

_____

_____

You will revisit your response to the Science Probe at the end of the lesson.

Distribution of Resources

Copyright © McGraw-Hill Education Pgiam/E+/Getty Images

ENCOUNTER
THE PHENOMENON

| Why are some areas rich in resources while others have so few?

Resources are not equally distributed in every part of the world. Some areas, like Saudi Arabia, are rich in energy resources. Other regions, like South Africa, are rich in mineral resources.

GO ONLINE

Check out *Natural Resources* to see this phenomenon in action.

Record your observations about the phenomenon in the space provided. Brainstorm with a partner reasons why this occurs.

EXPLAIN
THE PHENOMENON

Did you notice how mineral and energy resources are located in different regions around the world? Are you starting to get some ideas about why this occurs? Use your observations about the phenomenon to make a claim about why Earth's resources are unevenly distributed around the globe.

CLAIM

The uneven distribution of Earth's resources is the result of...

 COLLECT EVIDENCE as you work through the lesson. Then return to these pages to record your evidence.

EVIDENCE

A. What evidence have you discovered to explain why minerals only form in certain areas?

B. What evidence have you discovered to explain the uneven distribution of soils?

MORE EVIDENCE

C. What evidence have you discovered to explain why fossil fuels are unevenly distributed?

When you are finished with the lesson, review your evidence. If necessary, based on the evidence, revise your claim.

REVISED CLAIM

The uneven distribution of Earth's resources is the result of...

D. What evidence have you discovered to explain the geologic processes that influence the distribution of groundwater?

Finally, explain your reasoning for how and why your evidence supports your claim.

REASONING

The evidence I collected supports my claim because...

Where are minerals found on Earth?

The basic building blocks for soil, rocks, and metals are minerals. Metal products from airplanes to zippers come from minerals. The salt you put on your eggs at breakfast is a mineral. Even the exterior of the Statue of Liberty is coated in a mineral—copper! How do these important resources form and where are they found? Let's investigate!

INVESTIGATION

Location, Location, Location

The map below shows the location of copper deposits that are associated with igneous intrusions.

- • Copper deposit
- — Plate boundary

1. What patterns do you notice among the distribution of copper deposits?

_____

_____

_____

_____

_____

2. What evidence can you gather from your observations about the processes that form copper across the globe?

_____

_____

3. Based on the processes that you listed above, would you categorize copper as a renewable or nonrenewable resource? Explain your reasoning.

_____

_____

Hydrothermal Deposits Did you notice that many of Earth's copper deposits are concentrated near tectonic plate boundaries? Metallic minerals, including copper, gold, silver, lead, iron, and zinc, are associated with igneous intrusions and tied to plate tectonics. Intense heat from tectonic activity produces hot, mineral-rich fluids that chemically react with rocks. Minerals that crystallize from these fluids are called hydrothermal deposits. Some of these deposits closely correspond to **subduction zones**—areas where one tectonic place sinks beneath another. Other deposits represent very old intrusions that formed in the distant geologic past when the configuration of plate boundaries was different from today.

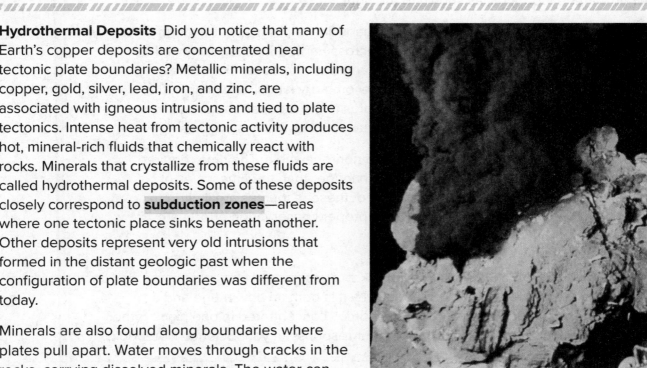

Minerals are also found along boundaries where plates pull apart. Water moves through cracks in the rocks, carrying dissolved minerals. The water can gush out of a hydrothermal vent, or opening, on the ocean floor. Minerals precipitate out of solution and are deposited around the vent. These minerals form large chimney structures as seen to the right.

▶ **GO ONLINE** for additional opportunities to explore!

Want to learn more about mineral formation? Then read one of the scientific texts.

☐ **Read** about the formation of diamonds in the **Scientific Text** _All About Diamonds._

OR

☐ **Read** about the relationship between minerals and mountains in the **Scientific Text** _Mountain Minerals._

Minerals from Cool Solutions Not all minerals are associated with plate tectonics. Sometimes minerals can crystallize as water evaporates. As water seeps into the ground or flows over Earth's surface, it interacts with minerals in rocks and the soil. The water dissolves some of these minerals and picks up elements such as potassium, calcium, iron, and silicon. These elements become dissolved solids. During dry conditions, as water evaporates, solids crystallize out of the water and form minerals. A deposit of the mineral halite—common rock salt—forms when water evaporates, as shown to the right.

Distribution of Minerals Because of the geoscience processes that form minerals, they are not distributed evenly across Earth's surface. Many of the deposits we mine on land were actually formed on the ocean floor. These rocks were uplifted from the seafloor to become dry land. For example, the copper mines located on the island of Cyprus in the Mediterranean Sea were formed by hydrothermal activity on the seafloor millions of years ago.

Mineral resources are typically limited and nonrenewable. The time required for minerals to form is, in most cases, much longer than the timescales of human lifetimes. Therefore, the quantities of these resources are limited to current and near-future generations if the proper conservation efforts are in place.

THREE-DIMENSIONAL THINKING

Construct an explanation about why it is both an advantage and a disadvantage to have mineral resources concentrated in one place rather than scattered all around. Write your response in your Science Notebook.

HISTORY ❭ Connection In 1974, a new mineral was discovered in the Dominican Republic and is now adding to the country's economy. Called larimar, this turquoise-colored mineral is found in only one locality in a fairly inaccessible region. Larimar is volcanic in nature, meaning that it formed as molten material that cooled and crystallized. Larimar is currently used in jewelry of all kinds.

COLLECT EVIDENCE

Why are the locations of minerals limited on Earth? Record your evidence (A) in the chart at the beginning of the lesson.

Ed Mathez works in mines such as this one near the Bushveld-Igneous Complex in South Africa. He examines layered igneous rocks in search of rare metals. ▶

Billions of Years in the Making

Ed Mathez investigates ore deposits to determine the source of rare metals.

Platinum—it is a precious metal, rarer than gold and silver. In fact, most of the world's platinum is mined from just five locations: Canada, Russia, South Africa, the United States, and Zimbabwe. To Ed Mathez, these sites contain more than just precious metals—they contain clues to Earth's past. Mathez, a geologist for the American Museum of Natural History in New York City, studies platinum ore deposits and the geologic processes that led to their formation long ago.

Mathez makes maps of layered igneous rocks in the Bushveld Igneous Complex, South Africa. Layers formed as magma, or molten rock beneath Earth's surface, cooled and crystallized within Earth's crust. The layers at the bottom of the Bushveld Igneous Complex contain high concentrations of dense sulfide deposits: iron, sulfur, and chromium-rich minerals. Sulfide deposits also contain tiny concentrations of platinum. Mathez has determined that the densest layers near the bottom of the formation hold an average of 6 to 8 parts per million (ppm) of platinum. That means for every 1 million atoms, only about 6 are platinum. It's a huge job to extract platinum from the rocks in order to produce profitable amounts of platinum. But it's worth the effort. Today, many important products and new technologies rely on this precious metal.

From Mine to Market

1 **Extraction** Miners extract the metal-rich rock from the ore deposit, crush it, and mix it with a liquid called froth. Metal sticks to the froth, and is skimmed off the top.

2 **Concentration** The froth is dried. The remains are melted in a furnace to separate the metals from the nonmetals.

3 **Refining** Platinum is separated from other metals, such as nickel and gold. Refinement requires chemical reactions to separate metals.

4 **Manufacturing** Platinum is typically mixed with other metals to produce jewelry, electronic equipment, and catalytic converters in automobiles.

It's Your Turn

STEM Careers Would you like to be a geologist? Write a paragraph describing what geologists do and whether or not you would want to be one. Share your reasoning with a partner.

▲ Platinum's durability and shiny surface make it a popular choice for jewelry. Platinum helps to convert toxic gas emissions into less harmful gases in catalytic converters.

Which locations have the most soil?

Have you ever grown a garden? If so, you have used one of the most important natural resources on Earth—soil. **Soil** is the loose, weathered material in which plants grow. How does this important resource form?

INVESTIGATION

Digging In

Research the processes involved in the formation of soil. In the space below, create an illustration that explains the formation of soil. Create descriptive captions to support each step.

Soil Formation If you dig down into ground, you would see that soil has a layered structure. At some point in your digging, you would likely strike solid rock. Soil forms directly on top of the rock from which it is made. In most areas it takes 80 to 400 years to form about 1 cm of topsoil. It begins when weathering by water, ice, and other agents cracks and breaks down rock. Plants, bacteria, and burrowing organisms continue the process of weathering. They help break down rock. If all soil forms in a similar process, why does soil differ from place to place?

It's a Slippery Slope

Study the landscape below.

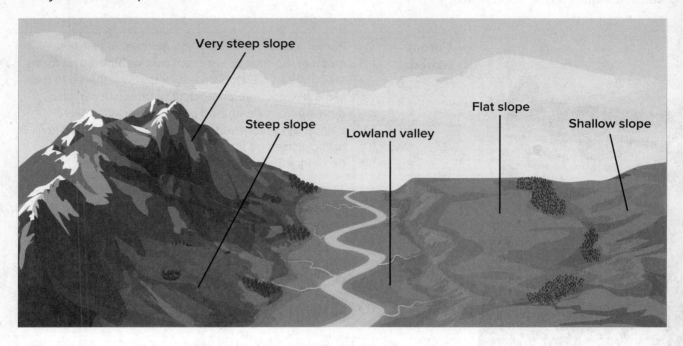

1. Hypothesize where you think soil develops the fastest. Explain your reasoning.

_____

_____

_____

_____

_____

_____

2. What conditions do you think might affect soil formation?

_____

_____

_____

_____

_____

Factors Affecting Soil Formation The quality and composition of soil vary from place to place, depending on how the soil was formed. The many kinds of soil that form depend on five factors. These factors include parent material, climate, topography, living things, and time. Examine the table below to learn more about each factor.

Parent Material	Parent material is the starting material of soil. It is made of the rock or sediment that weathers and forms soil. Soil can develop from rock that weathered in the same place where the rock first formed. It can also develop from weathered pieces of rock that were carried by wind or water from another location. The particle size and the type of parent material help determine the properties of the soil in an area.
Climate	Temperature and precipitation help determine an area's climate, or average weather. If the climate is warm and wet, soil formation can be rapid. But heavy rains carry away nutrients, so the soil is not good for growing plants. Rates of weathering tend to be low in dry climates and cold climates, so soils form slowly in these places. Areas with moderate temperatures and moderate amounts of precipitation tend to have rich soils.
Topography	Topography is the shape and steepness of the landscape. The topography of an area determines what happens to water that flows over the surface. In flat landscapes, most of the water enters the soil. In steep landscapes, most of the water flows downhill. It carries soil with it, leaving some slopes bare of soil. The soil is often deposited at the bottom of the slope. Here, soils tend to be thick.
Living Things	The organisms in soil range from tiny bacteria to furry moles. Living things help speed up the process of soil formation. They form passages for water to move through. When they decompose, they add organic matter to the soil.
Time	As time passes, weathering is constantly acting on rock and sediment. Soil formation is a slow, but steady process. Mature soils develop layers as new soil forms on top of older soil. Each layer has different characteristics as organic matter is added or as water carries elements and nutrients downward.

Copyright © McGraw-Hill Education (t)Photo by Lynn Betts, USDA Natural Resources Conservation Service, (b)photographyfirm/ Shutterstock.com

Warm, wet climates produce soil fastest. But large amounts of rain can wash away nutrients. The thickest, richest soils tend to be found in areas with moderate climates and gentle topography, where soils have been forming for a long period of time with little erosion.

THREE-DIMENSIONAL THINKING

Explain the role that geologic processes play in the formation of soil.

_____

_____

_____

_____

COLLECT EVIDENCE

What factors lead to an uneven distribution of soils across the globe? Record your evidence (B) in the chart at the beginning of the lesson.

Why are some regions rich in fossil fuels?

You might turn on a lamp to read, turn on a heater to stay warm, or ride the bus to school. In the United States, the energy to power lamps, heat houses, and run vehicles probably comes from nonrenewable energy resources, such as fossil fuels. Coal, oil, also called petroleum, and natural gas are fossil fuels.

Coal	Oil	Natural Gas

Read a Scientific Text

Coal, oil, and natural gas are important energy resources. Read the passage to learn more about formation processes and accessibility of coal, oil, and natural gas.

Inspect

Read the passage *Millions of Years Ago...*

Find Evidence

Reread the last two paragraphs. Underline how oil and natural gas are trapped.

Make Connections

Communicate With your partner, collect evidence from informational texts, such as encyclopedias or other relevant sources, to support the location of oil and natural gas traps in the United States. Plot these locations on a map. Compare and contrast your map with those of other students. Determine why any differences may have resulted.

Millions of Years Ago...

Coal Formation

Earth was very different 350 million years ago, when the coal used today began forming. Plants, such as ferns and trees, grew in prehistoric swamps. When plants in prehistoric swamps died, their remains built up. Over time, sediment covered the plant remains. Inland seas formed where the swamps once were. Bacteria, extreme temperatures, and pressure acted on the plant remains over time. Eventually a brownish material, called peat, formed. As additional layers of sediment covered and compacted the peat, over time it changed successively into harder types of coal.

Oil and Natural Gas Formation

Like coal, the oil and natural gas used today formed millions of years ago. The process that formed oil and natural gas is similar to the process that formed coal. However, oil and natural gas formation involves different types of organisms. Scientists theorize that oil and natural gas formed from the remains of microscopic marine organisms called plankton. The plankton died and fell to the ocean floor. There, layers of sediment buried their remains. Bacteria decomposed the organic matter, and then pressure and extreme temperatures acted on the sediments. During this process, thick, liquid oil formed first. If the temperature and pressure were great enough, natural gas formed.

Oil and natural gas can be trapped in the subsurface by various combinations of rock types and geologic structures. To trap oil or gas, there must be a rock in which these resources can accumulate. Such a rock unit is known as a reservoir.

Most of the oil and natural gas used today formed where forces within Earth folded and tilted thick rock layers. Often hundreds of meters of sediments and rock layers covered oil and natural gas. However, oil and natural gas were less dense than the surrounding sediments and rock. As a result, oil and natural gas began to rise to the surface by passing through the pores, or small holes, in rocks. Oil and natural gas eventually reached layers of rock through which they could not pass, or impermeable rock layers. Deposits of oil and natural gas formed under these impermeable rocks. The less-dense natural gas settled on top of the denser oil. Trapped oil can be discovered by geologists and extracted through drilling.

Geologic Traps Scientists theorize that oil and natural gas formed from the remains of plankton that lived in oceans. Yet oil deposits can be found beneath land as well as oceans. The reason oil can be found beneath land is because Earth's tectonic plates have shifted greatly over millions of years since the plankton existed. Let's examine how oil and natural gas get trapped.

INVESTIGATION

Between a Rock and a Hard Place

Analyze the image.

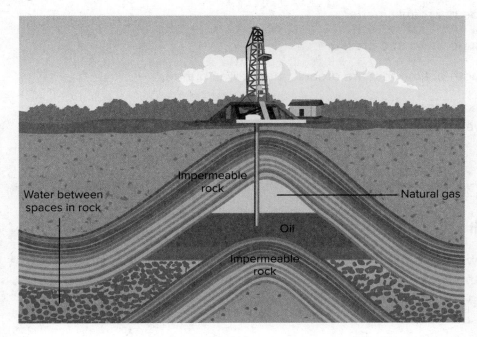

1. What feature prevents oil and natural gas from rising to the surface?

   _____

   _____

2. Explain how this feature allows for humans to access oil and natural gas reserves.

   _____

   _____

   _____

   _____

Want more information?
Go online to read more about why
Earth's resources are unevenly
distributed.

FOLDABLES
Go to the Foldables® library to make a
Foldable® that will help you take notes
while reading this lesson.

Fossil Fuel Formation The fossil fuels we use today formed from the
remains of prehistoric organisms. The decayed remains of these organisms
were buried under layers of sediment, then changed chemically by extreme
temperatures and pressure. The type of fossil fuel that formed—coal, oil, or
natural gas—depended on three factors:

- the type of organic matter,

- the temperature and pressure,

- the length of time that the organic matter was buried.

Given the conditions necessary for their formation, fossil fuels are not
distributed evenly around the world. Where are deposits of fossil fuels
located? Let's find out!

INVESTIGATION

Striking Oil

This map shows the distribution of the main sedimentary basins that contain
petroleum in the lower 48 states. Observe which parts of the country have
petroleum and which do not.

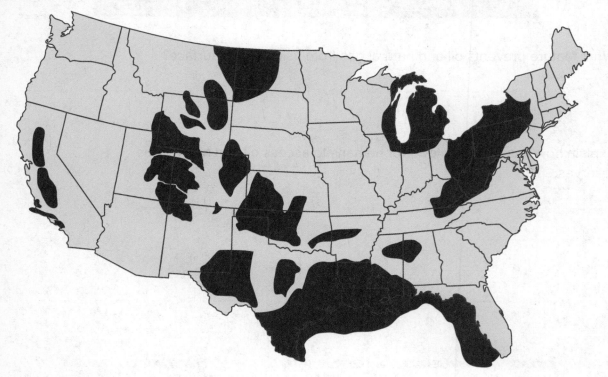

Copyright © McGraw-Hill Education

1. Briefly summarize where petroleum basins are located in the United States.

_____

_____

_____

_____

_____

2. Using your knowledge of how fossil fuels form, brainstorm why some regions have petroleum basins while others do not.

_____

_____

_____

_____

Location of Fossil Fuels The existence of fossil fuels in an area depends on the geologic history of the area. Places rich in coal were swamps hundreds of millions of years ago. Places rich in oil and natural gas were covered by ancient oceans. If these environmental conditions did not exist in an area, fossil fuels did not form. In the case of oil and natural gas, if these fuels were not sufficiently confined in geologic traps, they would not be available to extract.

According to the American Coal Foundation, each person in the United States uses approximately 3.8 tons of coal each year, enough to fill three pickup trucks! This amount includes the coal that is burned to provide heat and electricity, as well as the coal that is burned in manufacturing and other industrial processes.

THREE-DIMENSIONAL THINKING

Create a graphic organizer that **explains** how the geologic processes involved in the formation of coal, oil, and natural gas **cause** uneven distributions on Earth.

Why are these resources considered nonrenewable resources?

_____

_____

_____

_____

Fossil Fuels and Time Fossil fuel resources continue to be formed in the same ways that they were in the past. The amount of time required to form most of these resources is much longer than timescales of human lifetimes. Therefore, these resources are limited to current and near-future generations.

COLLECT EVIDENCE

Why are fossil fuels unevenly distributed on Earth? Record your evidence (C) in the chart at the beginning of the lesson.

Where is abundant groundwater located?

The freshwater beneath Earth's surface is much more plentiful than the freshwater in lakes and streams. Water that lies below ground, called groundwater, makes up about one-third of Earth's freshwater. Although Earth's crust appears solid, it is composed of soil, sediment, and rock that contain countless small openings, called pore spaces. What materials do you think have the most pore space? Let's explore!

INVESTIGATION

Fill the Void

Observe the sediment samples below.

Well-sorted

Poorly sorted

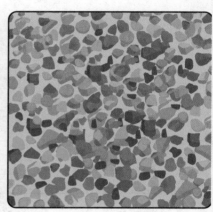
Well-sorted

1. What do you think the difference is between well-sorted sediment and poorly sorted sediment?

_____

_____

2. Which sample has the greatest porosity? Explain.

_____

_____

_____

_____

3. How do you think the process of cementation—when minerals crystallize between sediment grains—might affect the porosity of a material?

_____

_____

_____

4. **MATH Connection** Porosity is the fraction of material occupied by empty spaces. It can be expressed as a percent or a decimal. Assume that the porosity of a sample of sand and a sample of gravel is the same: 30 percent (0.30). Determine the porosity of a mixture of the sand and gravel, assuming that the sand fills the spaces between the gravel.

Porosity Pore spaces make up large portions of the materials that make up Earth's crust. The amount of pore space in a material is its **porosity.** The greater the porosity, the more water can be stored in the material. The porosity of well-sorted sediment—sediment that is all about the same size—is greater than that of poorly sorted sediment. In poorly sorted sediment, smaller particles occupy some of the pore spaces and reduce the overall porosity of the sediment. Similarly, the cement that binds the grains of sedimentary rocks together reduces the rocks' porosity.

Permeability The measure of water's ability to flow through sediment and rock is called **permeability.** This ability to flow through rock and sediment depends on pore size and the connections between the pores. Even if pore space is abundant in a rock, the pores must form connected pathways for water to flow easily through rock.

Limestone can have high porosity and permeability.

Copyright © McGraw-Hill Education PRILL/Shutterstock.com

Groundwater Storage People often bring groundwater to Earth's surface by drilling wells. Wells are usually drilled into an aquifer—an area of permeable sediment or rock that holds significant amounts of water. Let's investigate areas where groundwater deposits are found.

INVESTIGATION

Deeper Understandings

Examine the map below showing the topography of the United States.

1. Imagine you are a hydrologist looking for groundwater deposits. Identify areas that you think would contain the most groundwater resources by circling them on the map above.

2. Explain your reasoning for the areas you chose.

_____

_____

_____

_____

3. What processes do you think are responsible for the distribution of groundwater?

_____

_____

_____

Groundwater Distribution Groundwater supplies are limited based on distribution. They are the result of past and current geologic processes, including the water cycle. Recall that water is in constant circulation on Earth's surface, moving from ocean to atmosphere, from atmosphere back to the surface, and in and out of the subsurface of Earth. Processes in the water cycle, including precipitation, infiltration, groundwater flow, and the volume of groundwater naturally moved back to the surface, can all affect groundwater distribution. The locations of features on Earth's surface, such as mountains, have a strong impact on where precipitation occurs. Warm, moist air flowing up a mountain can cool, condense, and form rain along the windward side of the mountain.

The rock cycle also plays an important role in the distribution of groundwater. Surface drainage and the porosity and permeability of rocks below Earth's surface or sediment can influence where water collects underground. Some layers of rock, especially sedimentary rock, tend to be more porous than others and allow water to flow freely. The best groundwater basins are in valleys where a large amount of sediment has continuously been eroded and deposited. These valleys often contain porous, unconsolidated rocks. Plate motions typically determine the shape of these basins and are the cause of mountains being uplifted in the first place. The depth of impermeable rock, such as granite, can support or block large amounts of groundwater from collecting.

The Central Valley aquifer system is located in central California. Nearly all of the water received by this aquifer is provided by runoff from the mountains that surround the valley.

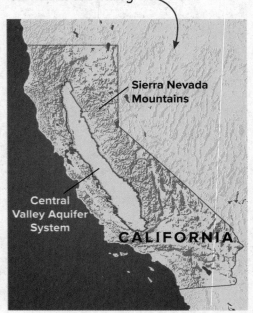

COLLECT EVIDENCE
What geologic processes influence groundwater distribution? Record your evidence (D) in the chart at the beginning of the lesson.

A Closer Look: Protecting Our Water Supply

Freshwater is Earth's most precious natural resource. Changes to groundwater supplies can lead to environmental issues such as overuse, subsidence, and pollution.

- **Overuse** Groundwater supplies can be depleted. If groundwater is pumped out at a rate greater than the recharge rate, the groundwater supply will decrease and the water table will drop.

- **Subsidence** When the height of the water table drops, the weight of the overlying material is increasingly transferred to the aquifer's mineral grains, which then squeeze together more tightly. As a result, the land surface above the aquifer sinks.

- **Pollution in Groundwater** Sources of groundwater pollution include sewage from faulty septic tanks and farms, landfills, and other waste disposal sites. Pollutants can spread rapidly through a highly permeable aquifer.

Groundwater Pollution Sources
Infiltration from fertilizers
Leaks from storage tanks
Drainage of acid from mines
Seepage from faulty septic tanks
Saltwater intrusion into aquifers near shorelines
Leaks from waste disposal sites
Radon gas from radioactive decay of uranium in rocks and sediment

There are a number of ways by which groundwater resources can be protected and restored. First, major pollution sources need to be located, identified, and eliminated. Pollution can enter groundwater resources via runoff and infiltration, or directly from underground. Pollution that already exists can be monitored with observation wells and other techniques. Most pollution spreads slowly providing adequate time for alternate water supplies to be found. In some cases, pollution can be stopped by building impermeable underground barriers around the polluted area. Sometimes, polluted groundwater can be pumped out for chemical treatment on the surface.

It's Your Turn

ENVIRONMENTAL〉Connection With a partner, discuss the best way to preserve the health of aquifer systems and prevent groundwater pollution in a residential area. Share your ideas with your class and work together to compile an informative blog post that can be shared with your community.

Summarize It!

1. **Organize** Complete the concept map to organize information you learned throughout the lesson about the geologic processes involved in the formation of minerals, soil, fossil fuels, and groundwater.

Minerals	Soil

Fossil Fuels	Groundwater

2. How does the formation of a resource affect its distribution?

_____

_____

_____

_____

Three-Dimensional Thinking

Use the map to answer the question.

United States Coal Reserves

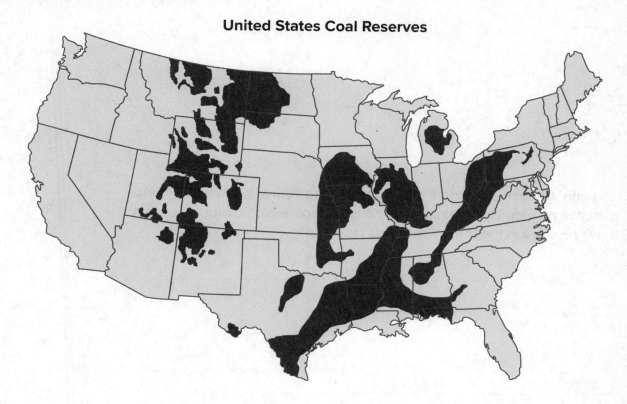

3. Coal deposits are unevenly distributed around the United States. Which explanation best supports the reasoning for why this occurs?

   A Some areas contain coal deposits because these areas were located near the South Pole millions of years ago. Over time, tectonic processes caused the continents to drift to their present locations, forming the coal we use today.

   B Some areas contain coal deposits because the conditions that form this energy resource are occurring in these places today. These regions are experiencing extreme temperatures and pressures, forming the coal we use today.

   C Some areas contain coal deposits because millions of years ago prehistoric swamps were covered in sediment. Bacteria, extreme temperatures, and pressure acted on the plant remains over time forming the coal we use today.

   D Some areas contain coal deposits because layers of sediment buried the remains of marine organisms millions of years ago. Bacteria decomposed the organic matter, and then pressure acted on the sediments forming the coal we use today.

Real-World Connection

4. Determine Why is the soil near your school different from the soil along a riverbank or the soil in a desert?

_____

_____

_____

_____

_____

5. Explain Choose a resource that is common to your state. What evidence can you provide to explain the past and current geologic processes that have resulted in the formation of this resource?

_____

_____

_____

_____

 Still have questions?
Go online to check your understanding about why Earth's resources are unevenly distributed.

REVISIT SCIENCE PROBES
Do you still agree with the statement you chose at the beginning of the lesson? Return to the Science Probe at the beginning of the lesson. Explain why you agree or disagree with that statement now.

EXPLAIN THE PHENOMENON

Revisit your claim about why some areas are rich in resources and others have so few. Review the evidence you collected. Explain how your evidence supports your claim.

KEEP PLANNING
STEM Module Project
Science Challenge

Now that you have learned about how resources are distributed based on past and present geologic processes, go to your Module Project to continue planning your news report. Keep in mind that you want to explain why some regions might be rich in resources and others are not.

Resource Extraction

Three friends were arguing about how the removal and use of resources by humans impacts the availability and distribution of resources. This is what they said:

Camila: I think the extraction and use of resources does not affect the amounts of these resources available on Earth, and does not change the overall distribution of these resources.

Caleb: I think removing and using resources decreases the amounts of these resources available in some locations, and changes the overall distribution of these resources on Earth.

Jasmine: I think the extraction and use of resources decreases the amounts of these resources available in some locations, but does not change the distribution of these resources on Earth.

Which friend do you agree with the most? Explain why you agree.

_____

_____

_____

_____

You will revisit your response to the Science Probe at the end of the lesson.

Depletion of Resources

ENCOUNTER
THE PHENOMENON | What is a mining "ghost town?"

When gold was discovered at Sutter's Mill in 1848, people from all over the world traveled to California in search of riches. As one official reported, "The farmers have thrown aside their plows, the lawyers their briefs, the doctors their pills, the priests their prayer books, and all are now digging gold." As people rushed to a new area to look for gold, they built new communities. Towns appeared almost overnight. One site on the Yuba River had only two houses in September, 1849. A year later, a miner arrived to find 1,000 people "with a large number of hotels, stores, groceries, [and] bakeries." When the gold rush ended, towns like the one shown here were abandoned.

With your partner, decide whether or not you think it's likely that abandoned mines and ghost towns could come back to "life." Why or why not? Record your thoughts in the space provided.

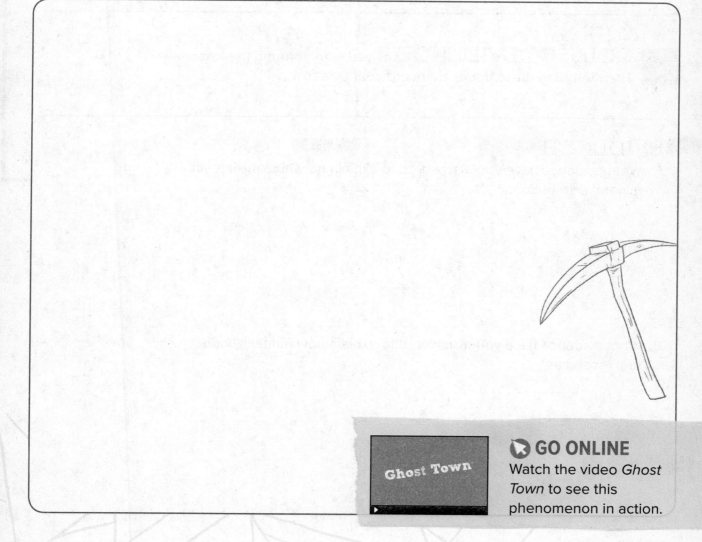

Ghost Town

GO ONLINE
Watch the video *Ghost Town* to see this phenomenon in action.

EXPLAIN
THE PHENOMENON

Are you starting to get some ideas about how the extraction and use of resources can change how much and where more of that resource can be found? Use your observations about the phenomenon to make a claim about how humans impact the distribution of resources on Earth.

CLAIM

Humans impact the distribution of resources by...

 COLLECT EVIDENCE as you work through the lesson.
Then return to these pages to record your evidence.

EVIDENCE

A. What evidence have you discovered to explain how humans impact mineral distributions?

B. What evidence have you discovered to explain how humans impact fossil fuel reserves?

MORE EVIDENCE

C. What evidence have you discovered to explain how humans impact groundwater resources?

When you are finished with the lesson, review your evidence. If necessary, based on the evidence, revise your claim.

REVISED CLAIM

Humans impact the distribution of resources by...

Finally, explain your reasoning for how and why your evidence supports your claim.

REASONING

The evidence I collected supports my claim because...

How does mining affect mineral distribution?

Recall that mineral deposits are not evenly distributed around the globe. As could be expected, the extraction and use of minerals further impacts how much and where more of that resource can be found. How do people extract different minerals?

Mining is the process by which commercially valuable resources are removed from Earth. These resources include ores, such as metals; precious stones, such as diamonds; and building stones, such as granite. In the following activity you will be a miner and simulate the extraction of minerals.

 Want more information?
Go online to read more about the depletion of mineral, energy, and groundwater resources.

FOLDABLES

Go to the Foldables® library to make a Foldable® that will help you take notes while reading this lesson.

LAB Mineral Mining

Safety

Materials

birdseed (700 g)
shallow pan
colored beads (30)

tweezers
paper towel
stopwatch

Procedure

1. Read and complete a lab safety form.

2. Divide into groups of 4–6 students. Pour the birdseed into the pan.

3. Add 30 colored beads to each pan.

4. "Mine" for the beads using tweezers for five minutes. Place mined beads on the paper towel. Any birdseed on the table or paper towel is considered a mining violation. Count and record the number of beads extracted per minute in the data table to the right.

5. Follow your teacher's instructions for proper cleanup.

Data and Observations	
Minute	Beads Extracted
1	
2	
3	
4	
5	

Analyze and Conclude

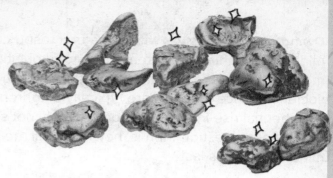

6. How difficult was it to mine the beads in the first minute?

_____

_____

7. How difficult was it to extract the beads in the last minute? Explain.

_____

_____

8. Plot the number of beads extracted during each minute on the grid below. Plot number of beads on the vertical axis and time on the horizontal axis.

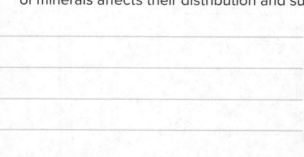

9. Using your graph, make a claim about how the extraction of minerals affects their distribution and supply.

_____

_____

_____

_____

Dwindling Deposits As with all natural resources, the demand for minerals continues to increase due to population growth and the fact that more countries are becoming industrialized and developing consumer societies. Since minerals are typically a nonrenewable resource, this exponential rise in consumption is simply not sustainable. An obvious question then is, how long will the world's mineral reserves last?

INVESTIGATION

Going, Going, Gone

Study the table below showing worldwide production rates and total reserves for some of the more common mineral resources. Then answer the questions that follow.

Mineral	Production (thousands of metric tons)	Reserves (thousands of metric tons)	Estimated Life of Reserves (years)
Iron ore	3,320,000	190,000,000	57
Aluminum ore (bauxite)	274,000	28,000,000	102
Phosphate rock	223,000	69,000,000	309
Chromium	27,000	>480,000	>18
Copper	18,700	720,000	39
Manganese	18,000	620,000	34
Zinc	13,400	200,000	15
Titanium concentrates	6,090	790,000	130
Lead	4,710	89,000	19
Nickel	2,530	79,000	31
Tin	294	4,800	16
Cobalt	124	7,100	57
Silver	27	570	21
Gold	3.0	56	19

Source: U.S. Geological Survey Mineral Commodity Summaries

1. What is the estimated life expectancy of tin reserves?

_____

_____

2. What is the estimated life expectancy of phosphate rock?

_____

_____

3. Why do you think the life expectancy of different mineral reserves varies?

_____

_____

COLLECT EVIDENCE

How does the removal of mineral resources affect how much and where more of these resources can be found? Record your evidence (A) in the chart at the beginning of the lesson.

Mineral Supplies For years scientists have warned about the prospect of important mineral reserves being depleted due to the exponential growth of demand. However, depletion has not occurred because total reserves have kept increasing. One reason for this is that geologists have improved their ability to locate new reserves through more sophisticated exploration techniques. Another factor is that technological advances in mining have allowed many low-grade deposits to become economical to extract. This does not mean that all mineral extraction can go on indefinitely. Even if additional supplies are found, the minerals will not be available to future generations if rates of consumption are not decreased.

How does the extraction of energy resources change their distributions?

From Earth's mineral resources we have made steel, concrete, and various materials that form the basis of our cities and factories, which, in turn, produce machines and countless types of products. What makes all of this possible are energy resources. Although there are different forms of energy, about 88% of the energy consumed by humans comes from the burning of fossil fuels. How are fossil fuels extracted?

LAB Coal Mining

Safety

Materials

salt dough
other materials provided by your teacher
ruler

Procedure

1. Read and complete a lab safety form.

2. Research the differences between strip-mining, underground mining, and mountaintop removal. Record your research in your Science Notebook.

3. Use the salt dough and other materials to build a model hill that contains coal deposits. Follow the instructions provided by your teacher on how to build the model.

4. Sketch the profile of the hill in the Data and Observations section on the next page. Use a ruler to measure the dimensions of the hill. Record the dimensions on the profile.

5. Decide which mining method to use to remove the coal. Mine the coal.

6. Try to restore the hill to its original size, shape, and forest cover.

7. Follow your teacher's instructions for proper cleanup.

Data and Observations

Analyze and Conclude

8. How does the mining of coal impact its overall distribution?

_____

_____

9. Compare the appearance of your restored hill to the drawing of the original hill. How does mining affect landscapes?

_____

_____

_____

_____

10. **ENVIRONMENTAL** **Connection** Describe two potential consequences of the lost forest cover and loose soil on the mined hill. What effect might mining have on the health of people and the natural environment?

_____

_____

_____

_____

_____

Fossil Fuel Extraction As you just explored in the *Coal Mining* lab, coal can be mined much like a mineral. Deposits of natural gas and oil can be extracted by drilling down into the ground. The deposits are often trapped between layers of impermeable rock. However, mining disturbs habitats and changes the landscape. If proper regulations are not followed, water can be polluted by runoff that contains heavy metals from mines.

Like minerals, fossil fuels are not distributed evenly around the globe and are nonrenewable. No one knows for sure when supplies will be gone. Let's take a closer look.

INVESTIGATION

Farewell Fossil Fuels

The chart shown here is based on data gathered by the U.S. Energy Information Administration (EIA). This agency analyzes information from around the world. Study the chart. Then answer the questions that follow.

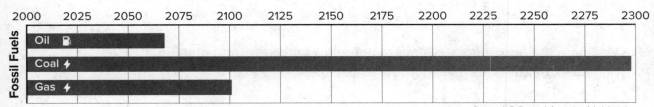

Source: U.S. Energy Information Administration

1. About how long is the world's oil expected to last? What about natural gas?

_____

_____

2. Coal reserves are projected to last until what year?

_____

_____

3. What might explain why these projections vary?

_____

_____

_____

_____

Supply and Demand As you might have gathered in the *Farewell Fossil Fuels?* investigation, coal is more abundant than oil or natural gas. Reserves of coal are projected to last more than 275 years, but natural gas and oil reserves may run out much sooner. Fossil fuels are forming all the time. However, we use them much more quickly than nature replaces them. They will be depleted unless more deposits are found, technology improves, or rates of usage change.

THREE-DIMENSIONAL THINKING

1. In the space below, create a graphic organizer showing the **cause-and-effect** relationships between humans and the removal of fossil fuels.

2. How does understanding these relationships enable you to make predictions about the future of fossil fuels?

_____

_____

_____

COLLECT EVIDENCE

How does the removal of energy resources affect how much and where more of these resources can be found? Record your evidence (B) in the chart at the beginning of the lesson.

How do humans impact groundwater resources?

Groundwater is the primary source of water for more than two billion people worldwide. Nearly half of all Americans depend on groundwater for drinking and other domestic purposes. In the following activity, you will investigate how the extraction and use of groundwater resources changes the availability and overall distribution of groundwater on Earth.

INVESTIGATION

Out of Sight, Out of Mind

Study the graph below showing changes in groundwater storage for seven major aquifer systems based on data collected by the *Gravity Recovery and Climate Experiment (GRACE)*.

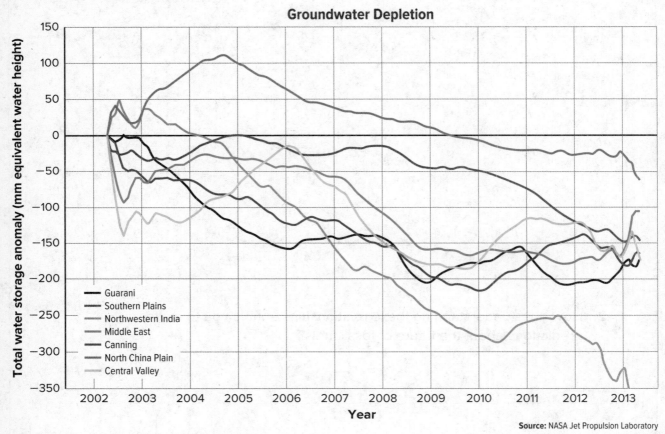

Source: NASA Jet Propulsion Laboratory

1. What trends or patterns do you notice in the data?

_____

_____

_____

GO ONLINE Now, watch the video *Irrigation and Groundwater Depletion*. Then answer the questions that follow.

2. What was the percent increase of groundwater depletion caused by irrigation between 2000 and 2011?

_____

_____

3. What do scientists use to help understand and predict where groundwater depletion is the most severe?

_____

_____

4. **ENVIRONMENTAL ⟩ Connection** Examine the figure below. Where is groundwater depletion predicted to be the most severe in 2082? What is the potential future risk for the human population if overexploited aquifers run dry? Can you think of another risk? Write your response in your Science Notebook.

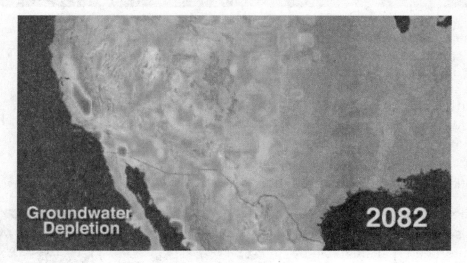

Groundwater Depletion In many areas of the world, groundwater is being withdrawn from aquifers faster than natural recharge can replace it. This is called **groundwater overdraft.** Excessive pumping for irrigation and other uses has removed so much water that wells have dried up in many places, and farms, ranches, and even whole towns are being abandoned.

COLLECT EVIDENCE

How does the removal of groundwater resources impact how much and where more of these resources can be found? Record your evidence (C) in the chart at the beginning of the lesson.

A Closer Look: GRACE

It may seem hard to believe that satellites orbiting Earth can "see" water below Earth's surface. However NASA's *Gravity Recovery and Climate Experiment (GRACE)* twin satellites did just that.

Using variations in Earth's gravity, *GRACE* was able to monitor the change in volume of underground aquifers. This data is particularly important as many of the world's major aquifers distributed around the globe are under strain due to expanding irrigation operations, urbanization, mining operations that need substantial water inputs, and droughts.

GRACE was the first tool that enabled scientists to visualize where people are using more water than is sustainable by directly measuring groundwater changes from space. Understanding changes in global water resources improves environmental monitoring and forecasting.

It's Your Turn

Copyright © McGraw-Hill Education (bkgd)NASA/JPL-Caltech, (inset)Justin Sullivan/Getty Images News/Getty Images, (b)Felix Landerer, NASA/Jr

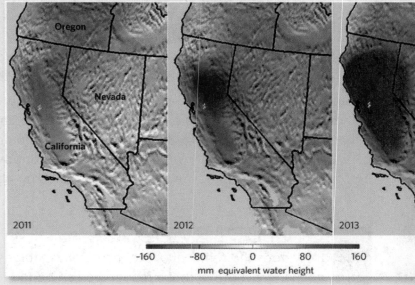

PHYSICAL SCIENCE ⟩ Connection

How did *GRACE* measure variations in Earth's gravity? Research and report your findings to the class.

Read a Scientific Text

You have learned that the extraction of groundwater can deplete this important resource, but did you know that overpumping wells can also cause water quality issues?

The quality, quantity, and reliability of groundwater resources are directly affected by the health of the aquifers. When wells are overdrawn, underlying salt water can rise into the wells and contaminate freshwater aquifers.

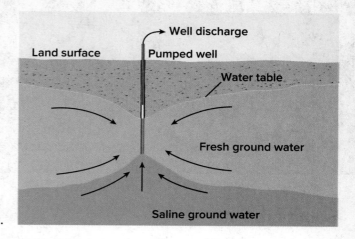

Land surface · Well discharge · Pumped well · Water table · Fresh ground water · Saline ground water

CLOSE READING

Inspect

Read the passage *Saltwater Intrusion.*

Find Evidence

Reread the second paragraph. Underline cause-and-effect relationships among events.

Make Connections

Communicate With your partner, identify the regions that are most at risk for salt water contamination. What is the potential for saltwater intrusion in your region?

Copyright © McGraw-Hill Education   TEXT CREDIT: "Saltwater Intrusion. USGS Online Publications Directory. https://pubs.usgs.gov/circ/circ1186/ pdf/p64-70.pdf

PRIMARY SOURCE

Saltwater Intrusion

The fresh ground-water resource of the United States is surrounded laterally and below by saline water. This is most evident along coastal areas where the fresh ground-water system comes into contact with the oceans, but it is also true in much of the interior of the country where deep saline water underlies the freshwater. The fresh ground-water resource being surrounded by saltwater is significant because, under some circumstances, the saltwater can move (or intrude) into the fresh ground-water system, making the water unpotable.

Freshwater is less dense than saline water and tends to flow on top of the surrounding or underlying saline ground water. Under natural conditions, the boundary between freshwater and saltwater maintains a stable equilibrium ... When water is pumped from an aquifer that contains or is near saline ground water, the saltwater/freshwater boundary will move in response to this pumping. That is, any pumpage will cause some movement in the boundary between the freshwater and the surrounding saltwater. If the boundary moves far enough, some wells become saline, thus contaminating the water supply. The location and magnitude of the ground-water withdrawals with respect to the location of the saltwater determines how quickly and by how much the saltwater moves. Even if the lateral regional movement of saltwater is negligible, individual wells located near the saltwater/ freshwater boundary can become saline as a result of significant local drawdowns that cause underlying saltwater to "upcone" into the well.

Source: United States Geological Survey

ENVIRONMENTAL Connection How can the overpumping of wells impact the health of freshwater aquifers, which in turn affects the quantity and quality of groundwater resources? In your Science Notebook, create a cause-and-effect graphic organizer to summarize your understanding.

Summarize It!

1. **Organize** your understanding of how humans impact the distribution of Earth's mineral, energy, and groundwater resources using the chart below. The topic is *Depletion of Resources*. Choose three central ideas you learned about in this lesson and provide specific details about each main idea.

Topic: _____

Main Ideas	Specific Details

Three-Dimensional Thinking

Analyze the graph showing groundwater storage changes in the Central Valley.

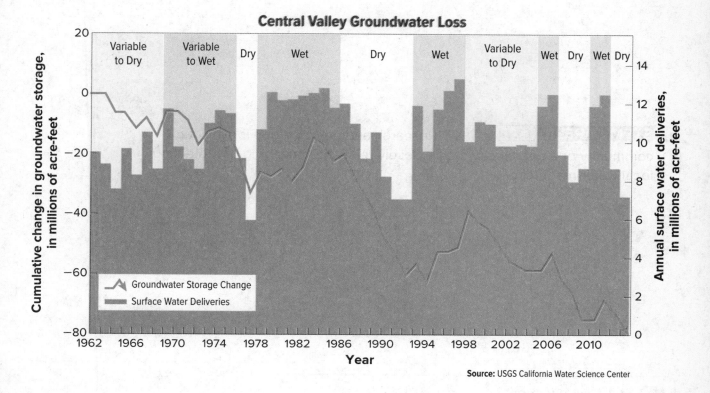

Central Valley Groundwater Loss

Groundwater Storage Change
Surface Water Deliveries

Source: USGS California Water Science Center

2. Which statement is a logical conclusion that can be drawn from information in the graph?

   A The increase in groundwater availability between 1962 and 2014 in the Central Valley can be explained by annual surface water deliveries.

   B The decrease in groundwater availability between 1962 and 2014 in the Central Valley can be explained by annual surface water deliveries.

   C The increase in groundwater availability between 1962 and 2014 in the Central Valley can NOT be explained by annual surface water deliveries.

   D The decrease in groundwater availability between 1962 and 2014 in the Central Valley can NOT be explained by annual surface water deliveries.

Real-World Connection

3. **Propose** a solution to the future depletion of mineral resources and energy resources. How might humans extend the life of these reserves?

_____

_____

_____

_____

4. **STEM CAREER ▸ Connection** Research and describe one job in your community or a nearby city that is closely related to providing or protecting local water resources.

_____

_____

_____

_____

_____

> **Still have questions?**
> Go online to check your understanding about the depletion of Earth's resources.

Copyright © McGraw-Hill Education Hans Debruyne/Shutterstock.com

REVISIT  SCIENCE PROBES

Do you still agree with the statement you chose at the beginning of the lesson? Return to the Science Probe at the beginning of the lesson. Explain why you agree or disagree with that statement now.

EXPLAIN THE PHENOMENON

Revisit your claim about how the removal of a natural resource impacts how much and where more of that resource can be found. Review the evidence you collected. Explain how your evidence supports your claim.

PLAN AND PRESENT

STEM Module Project
Science Challenge

Now that you've learned about depletion of Earth's resources, go back to your Module Project to continue planning your news report, construct your script, and present your newscast. Your goal is to explain why resources, such as salt, are not always readily available.

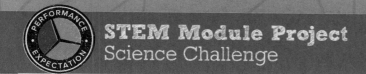
Where in the world…?

You are watching a local news report about how an unusually cold winter has brought freezing temperatures and icy conditions to several western states. The report states that rock salt, a common remedy for slippery roads, is not readily available in these states and is only found in certain parts of the world. Why do some resources only exist in certain areas?

You decide to conduct an investigation to find out 1) how people depend on Earth's resources, 2) why some regions on Earth are rich in certain resources while others are not, and 3) how people affect resource distributions. You'll use your findings to prepare a script that will be read on the 6 o'clock news explaining why resources, such as salt, are not always readily available.

Planning After Lesson 1

Brainstorm a list of different types of mineral, energy, and groundwater resources. Explain whether these resources are considered renewable or nonrenewable.

As a group, choose a mineral, energy, and groundwater resource to investigate. List these resources below. In your Science Notebook, describe how people are dependent upon the resources you selected.

Planning After Lesson 2

Which geologic processes are involved in the formation of your resources?
Describe or illustrate the processes below. Include evidence for any past or
current geologic processes related to the formation of your resources.

Gather data about the distributions of your chosen resources in your
Science Notebook. Sketch a world map below. Plot the locations of your
resources on the map.

Planning After Lesson 3

How has the extraction and use of your team's mineral, energy, and groundwater resources affected their distributions?

How have usage rates of the resources changed over time?

How will this affect resource availability for current and near-future generations? Be sure to consider factors that deplete the resources as well as factors that replenish the resources.

Prepare Your Newscast

Use the evidence you have gathered to write a script for your newscast explaining the causes of the uneven distribution of Earth's resources. Write your script in the space below. Use reasoning to connect the evidence you gathered and to support your explanation.

Evaluate Your Newscast

Use the criteria below to evaluate your scientific explanation and to describe any revisions you will make.

Does your scientific explanation...	Yes	No	Describe Revisions to Meet Criteria
...include the effects of past and present geologic processes?			
...include the effects of human activities?			
...use multiple valid and reliable sources of evidence to support your explanation?			

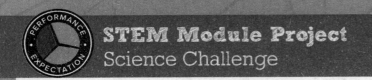

Be a News Anchor!

Present your scientific explanation before the class.

Take careful notes as each news anchor presents their newscast about the distribution of its resources. Then use your notes and the knowledge you've gained about your own resources to answer the questions below.

Compare and contrast distribution patterns for the resources researched by different groups. What overall factors affect resource distribution?

Do you think resources may be found in new places in the future? Why or why not? (Hint: *Assume that processes in the natural world will continue to operate in the future as they do today.*)

Humans depend on Earth's land, ocean, atmosphere, and biosphere for many different resources, including resources such as oil. What might happen if our supply of oil ran out? Consider both short-term and long-term consequences. Consider positive and negative impacts on people and the environment.

Congratulations! You've completed the Science Challenge requirements!

Module Wrap-Up

REVISIT
THE PHENOMENON

Using the concepts you've learned in the module, explain how the uneven distribution of Earth's mineral, energy, and groundwater resources are the result of past and current geoscience processes.

OPEN INQUIRY

What are one or two questions you still have about the phenomenon?

_____

_____

_____

Choose the question you are most interested in. Plan and conduct an investigation to answer this question.

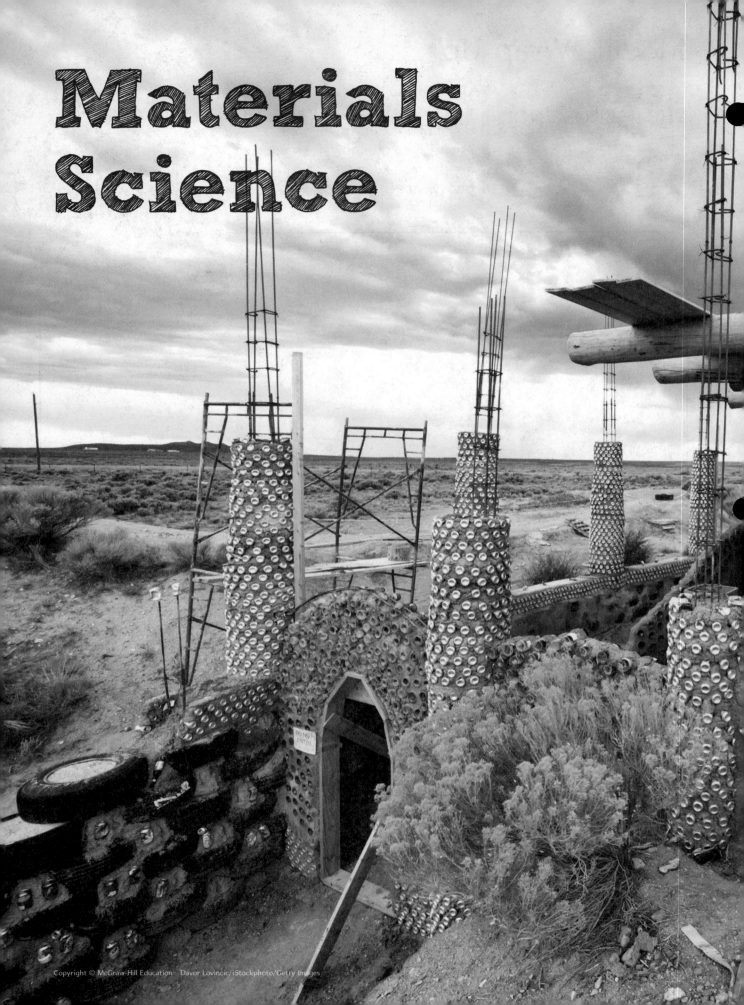

Materials Science

ENCOUNTER
THE PHENOMENON

What is this structure made of, and is it eco-friendly?

Houses of the Future

GO ONLINE
Watch the video *Houses of the Future* to see this phenomenon in action.

Communicate Think about what you know about the materials that make a shelter. With a partner, discuss your thoughts. Then record what you both would like to share with the class in the space below.

Take Cover

You work for a non-profit company that supplies shelters to people affected by natural disasters. Your company has asked you to be on a team that will develop a shelter that is economical and has a low impact on the environment.

You are tasked with researching what current technologies are used to make shelters and evaluating how well they meet the criteria. You will then design your own shelter and test whether it can withstand a natural disaster.

Start Thinking About It

How have people's needs and wants changed the features of shelters? In the image above, you can see the structure and function of the shelter. Identify any desired features the shelter has. Identify any desired features not pictured that you will incorporate into your design. Discuss your thoughts with your group.

Lesson 1
Synthetic Technology

Lesson 2
Impacts of Synthetic Materials

STEM Module Project

Planning and Completing the Engineering Challenge How will you meet this goal? The concepts you will learn throughout this module will help you plan and complete the Engineering Challenge. Just follow the prompts at the end of each lesson!

Recycled Materials

George and his friends were walking in the park when they spotted a new bench. The bench had a recycling symbol on the back. George said he had heard of this type of bench and claimed it was made from old milk jugs.

Angelica: I disagree with George. A milk jug can be squeezed and is somewhat flexible. The bench is not flexible.

Alex: I agree with George. The milk jugs went through a chemical process to change them.

Thomas: I disagree with George. The bench isn't the same color as milk jugs.

Mariah: I agree with George. I think the manufacturer stacked lots of milk jugs together until the jugs attached to each other.

Circle the student you agree with most. Explain your choice.

_____

_____

_____

_____

_____

_____

You will revisit your response to the Science Probe at the end of the lesson.

Synthetic Technology

ENCOUNTER
THE PHENOMENON | Why is polypropylene used to make drinking straws?

Take a look at the materials your teacher has provided. Upon first glance, you may think they are all the same. While they share some properties, they are not identical. Think critically about what makes each material unique. Record your observations and thoughts below.

Copyright © McGraw-Hill Education

Some
Assembly
Required
▶

◗ GO ONLINE
Watch *Some Assembly Required* to see this phenomenon in action.

EXPLAIN
THE PHENOMENON

You were probably familiar with many of the materials you observed in the activity. After taking a closer look, did you get any ideas about how the type of plastic determined what the plastic was used for? Make a claim about how the properties of a material relate to the function of that material.

CLAIM

The properties determine the function of a material because...

 COLLECT EVIDENCE as you work through the lesson. Then return to these pages to record your evidence.

EVIDENCE

A. What evidence have you discovered to explain how the properties of a material determine the function of the material?

B. What evidence have you discovered to explain the differences between natural materials and synthetic materials?

MORE EVIDENCE

C. What evidence have you discovered to explain how the formation of synthetic materials relates to its properties?

When you are finished with the lesson, review your evidence. If necessary, based on the evidence, revise your claim.

REVISED CLAIM

Synthetic materials are made from different materials because...

Finally, explain your reasoning for how and why your evidence supports your claim.

REASONING

The evidence I collected supports my claim because...

What are the properties of materials?

All of the plastics you looked at were made up of different materials. A **material** is the matter from which a substance is or can be made. Different materials have different properties based on the substances they are made from. Can you use the properties of a material to determine how a specific material is best used? Let's investigate!

 Want more information?
Go online to read more about synthetic materials.

FOLDABLES
Go to the Foldables® library to make a Foldable® that will help you take notes while reading this lesson.

Safety

Materials

water

magnifying lens

balance

graduated cylinder

materials for testing

Procedure

1. Read and complete a lab safety form.

2. Observe each material and test the material's hardness, flexibility, and reflectivity. Record your results in the data table in the Data and Observations section on the next page.

3. Using the balance, find the mass of each material. Record the mass.

4. Using the graduated cylinder, find the volume of each material using the displacement method. Record the volume.

5. While your material is in the graduated cylinder, note if it is absorbent or not.

6. Calculate the density of each material and record the values in the table.

7. Record any additional observations you make about the materials.

8. Follow your teacher's instructions for proper cleanup.

Data and Observations

Material	1	2	3	4	5
Hardness (hard or brittle)					
Flexible (yes or no)					
Reflectivity (shiny or dull)					
Mass					
Volume					
Density					
Absorbent (yes or no)					
Other observations					

Analyze and Conclude

9. What are some similarities and some differences among the materials? Are there any patterns in the properties?

_____

_____

_____

10. Choose one material. What could you use the material for? Based on its properties, why would it be well-suited for that use?

_____

_____

_____

Material Properties Each material has characteristic physical and chemical properties that can be used to identify it. Based on those properties, a material can be described in a variety of ways. For example, a material may be strong and flexible or absorbent and soft. The properties of a material can determine its behavior and how it is used. A material's properties can change according to how the material is treated or the conditions under which it is stored.

THREE-DIMENSIONAL THINKING

Suppose you have an unknown material that has the following properties: waterproof, rigid, and hard. The material is also brittle. Light passes easily through this material. Based on those properties, predict what the **function** of the material might be.

_____

_____

_____

_____

Copyright © McGraw-Hill Education

COLLECT EVIDENCE

How does knowing the properties of a material help explain how you could determine the function of the material? Record your evidence (A) in the chart at the beginning of the lesson.

What are the differences between natural and synthetic materials?

If there were no humans on Earth, would there be plastic? What about wood, metal, or glass? Some of these materials are natural materials. A **natural material** is any physical matter that is obtained or made from plants, animals, or the ground. Natural materials come from the natural environment and have undergone very little modification. Plastic is a synthetic material. A **synthetic material** is a material obtained from a natural material that has undergone a chemical reaction in a laboratory or factory. Both natural and synthetic materials play important roles in many of the products you use on a daily basis.

Which materials are natural and which are synthetic?

INVESTIGATION

Take Your Places

1. **WRITING Connection** Return to the Data and Observations table in the Lab *Materials Fair*. Identify the materials that are synthetic. Research what natural resources those synthetic materials came from. Obtain your information from at least two published sources. Record your findings below.

2. Now, create a classification system that explains how to organize materials based on where they came from. Consider the properties of the synthetic materials that make them different from the natural resources. Consider also how those properties contribute to the functions of the synthetic materials.

3. Examine the materials below. Add these materials to your classification system above.

Developing Synthetic Materials A materials scientist looks at the connections between the structure of a material and its properties to determine if a material can be changed or improved. He or she studies the desirable and undesirable properties of materials. Inspiration for synthetic materials often comes from natural materials. For example, a group of scientists are currently looking at sea cucumbers and how they can go from soft and squishy to rigid and tough when threatened. Materials scientists are trying to mimic this behavior by creating a fabric that will transition from flexible to inflexible. Any new synthetic material is considered a **technology**—the practical use of scientific knowledge.

Materials scientists test many different materials for useful properties.

Sources of Synthetic Materials Recall all matter must come from somewhere. Scientists must use an existing material to make any new synthetic material. The materials used to make synthetic materials come from natural resources. These natural resources must be obtained from plants, animals, or Earth.

COLLECT EVIDENCE

How can you explain some of the differences between natural materials and synthetic materials? Record your evidence (B) in the chart at the beginning of the lesson.

How are synthetic materials formed?

You know that synthetic materials come from natural resources and that inspiration for new synthetic materials comes from nature. But how do synthetic materials, such as the plastics at the beginning of the lesson, get created? Try your hand at making a synthetic material and see if you can figure it out!

 Slime Time

Safety

Materials

white school glue	plastic cup	balance
graduated cylinder	baking soda	saline solution
food coloring		

Procedure

1. Read and complete a lab safety form.

2. Pour 100 mL of white school glue into the plastic cup.

Procedure, continued

3. Add 7.5 g of baking soda and mix until combined.

4. Add 5 drops of food coloring and mix again.

5. Add 15 mL of saline solution and mix until the substance is firm.

6. Remove the substance from the cup and knead it until the mixture is soft.

7. Record your observations in the Data and Observations section below.

8. Follow your teacher's instructions for proper cleanup.

Data and Observations

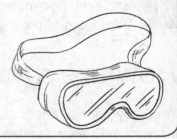

Analyze and Conclude

9. Compare the properties of the starting materials and the properties of the end product.

_____

_____

_____

10. What had to happen for the product to form?

_____

11. Based on your classification system, what synthetic material do you think you have made?

_____

_____

_____

New Materials Throughout history, humans have modified natural materials to enhance or improve their existing properties. This modification of natural materials creates new ones that do not exist naturally. Natural materials undergo chemical reactions in which the atoms of the natural materials rearrange and form new molecules. These molecules form the synthetic material.

INVESTIGATION

From This, To That

▶ **GO ONLINE** Watch the animation *How Paper Is Made*. Summarize the process that takes place to make trees into paper.

How does this, become this?

From Reactants to Products As you saw in the Lab *Slime Time* and in the Investigation *From This, To That,* all synthetic materials are the result of **chemical reactions.** Substances react chemically in characteristic ways. You may recall that in a chemical reaction, the bonds between atoms in the reactants are broken. The atoms rearrange and make new bonds to form the products.

The products have different properties than the reactants. For example, when you made the new material in the Lab *Slime Time*, the reactants underwent a chemical reaction called polymerization. Polymerization is the chemical process in which small organic molecules, or monomers, bond together to form a chain. A monomer is one of the small organic molecules that make up the long chain of a polymer. Polymer chains can be very long. Polypropylene, which is used to make drinking straws, might have 50,000 to 200,000 monomers in its chain.

THREE-DIMENSIONAL THINKING

Limestone, $CaCO_3$, is a natural material that can form from the remains of coral and the shells of marine organisms. For centuries, humans have been burning this material in kilns to produce calcium oxide, CaO, and a gas. When combined with water, sand, and small stones, calcium oxide becomes concrete. **Develop a model** in the space below to represent the limestone reacting to become calcium oxide. Using your understanding of **matter,** determine which gas is produced by the reaction.

Characteristics of Synthetic Materials Do you recall how the plastics were different from each other at the beginning of the lesson? Synthetic materials can come in a variety of different types. Examples of synthetic materials include plastics, ceramics, composites, synthetic fibers, artificial foods, medicines, or alternative fuels just to name a few. All synthetic materials share common characteristics. They are all made from natural resources, they all underwent a chemical reaction, and they are all made to fit a specific function based on the properties of the material. Look at the table below for some examples.

Synthetic Material	Properties	Examples
Synthetic fibers	flexible, can be spun	rayon, polyester, nylon
Ceramics	strong but brittle, good insulators	cement, tiles, bone china
Polymers	strong and flexible, easily modified to hold different shapes and colors	polyester, nylon, acrylic
Foods and Medicines	made to fix a specific function; mimics natural materials	vitamin C, red 40, hydrogenated oils
Composites	emphasize specific properties such as flexibility and strength	concrete, fiberglass

COLLECT EVIDENCE

How are synthetic materials formed and how does this relate to their properties? Record your evidence (C) in the chart at the beginning of the lesson.

A Day in the Life of a Materials Scientist

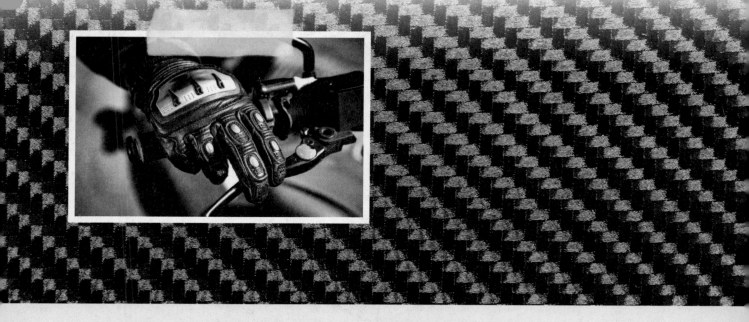

In 1964, Stephanie Kwolek's research group was assigned to create a strong, rigid, heat-resistant and lightweight polymer fiber. When Kwolek could not melt one polymer, she searched for a solvent that could dissolve the polymer. After many tries, Kwolek finally succeeded in dissolving the polymer, only to reach an unexpected result. The solution was watery and cloudy—different from any previous polymer.

Kwolek had the polymer spun into fibers and the results were astonishing. The fibers lined up parallel to each other in an orderly way creating a strong, lightweight, durable material. The group immediately recognized possible uses for such a material. The material is an aramid fiber, which is a group of synthetic materials classified by strength and heat resistance.

Aramid fiber is difficult to cut and does not shrink when exposed to cold temperatures. It will not rust, corrode or react with water. Aramid fiber is flame resistant and will not melt or soften at high temperatures. While Aramid fiber is best known for its use in bulletproof vests, it is also used in hiking boots, tires, gloves, brake pads, and hundreds of other products.

It's Your Turn

Organize Choose one material with which a materials scientist may work. Research the products or applications and the physical or chemical properties of that material. In your research, use at least two different sources including text, media, visual displays, or data. Share what you discover with your class.

Review

Summarize It!

1. **Outline** Complete the graphic organizer with the information you learned about natural and synthetic materials.

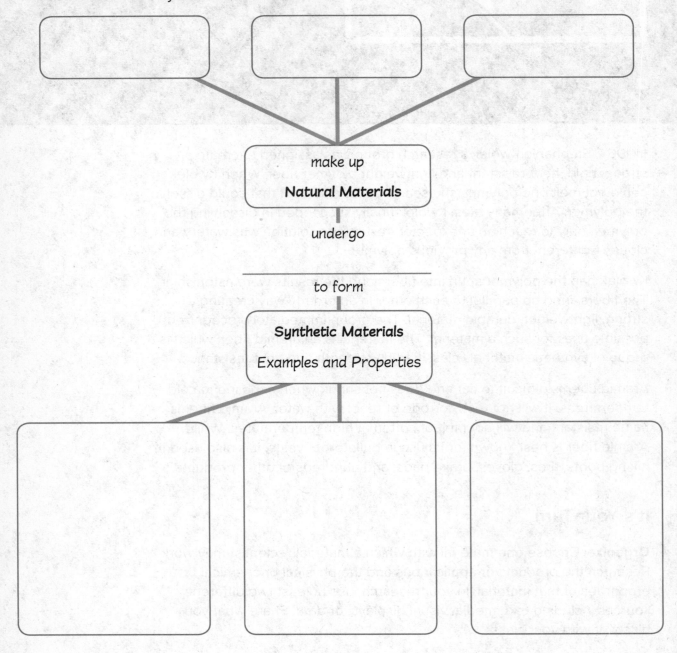

make up
Natural Materials

undergo

_____

to form

Synthetic Materials
Examples and Properties

Three-Dimensional Thinking

Tensile strength is a measure of the amount of "pulling" stress an object can withstand before it breaks or becomes damaged. The graph below shows the tensile strength for four materials being considered for a new product.

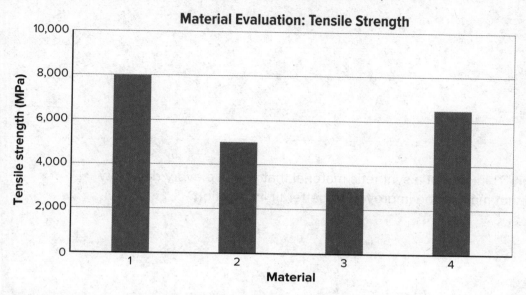

Material Evaluation: Tensile Strength

2. Which material should be considered if the product must be tear-resistant?

   A 1

   B 2

   C 3

   D 4

As an engineer working on the design of a new car, you need to select the right ceramic material to build parts of the car's engine and its onboard computer. The table below shows the materials you have to choose from.

Material	Wear Resistant	Conducts Electricity	Reacts with Chemical	Melting Point (°C)
1	highly	no	no	3,000
2	not at all	no	yes	100
3	moderately	yes	no	1,500
4	resistant	yes	no	500

3. Which of the above materials would you use when you build the engine?

   A 1

   B 2

   C 3

   D 4

Real-World Connection

4. Explain You are designing a new material for use in an airplane body. What properties should the material have? Explain.

_____

_____

_____

_____

_____

_____

5. Design Think about a synthetic material that you use every day. How could that material be improved to better fit its function?

_____

_____

_____

_____

 Still have questions?
Go online to check your understanding about synthetic materials.

REVISIT SCIENCE PROBES Do you still agree with the student you chose at the beginning of the lesson? Return to the Science Probe at the beginning of the lesson. Explain why you agree or disagree with that student now.

EXPLAIN THE PHENOMENON

 Revisit your claim about how properties determine the function of a material. Review the evidence you collected. Explain how your evidence supports your claim.

START PLANNING STEM Module Project Engineering Challenge

Now that you understand the differences between natural and synthetic materials, go to your Module Project and identify the properties that your shelter will need. Keep in mind the materials you could use to keep the structure eco-friendly.

Impacts of Synthetic Materials

Four friends were arguing about the impacts of synthetic materials. They each had different ideas about how synthetic materials impact our world. This is what they said:

Charlie: I think synthetic materials have negative impacts on the environment.

Dani: I think synthetic materials have positive impacts for individual people.

Obi: I think synthetic materials have positive and negative impacts for the environment and individual people.

Mei: I think synthetic materials have positive and negative impacts for individual people, but not the environment.

Circle the name of the friend you think has the best idea about the impacts of synthetic materials. Explain your choice.

_____

_____

_____

_____

_____

You will revisit your response to the Science Probe at the end of the lesson.

Synthetic Materials and Societal Impacts

ENCOUNTER
THE PHENOMENON
What are the impacts of making, using, and disposing of cell phones?

▶ GO ONLINE

Check out *Where Cell Phones Come From* to see this phenomenon in action.

Cell phones are used by millions of people around the world. Create a list of ways you think cell phones impact society and the environment.

EXPLAIN
THE PHENOMENON

Cell phones are made up of many different synthetic materials that come from many different natural resources. Make a claim about how synthetic materials, such as the ones used to make cell phones, impact society and the environment.

CLAIM

Synthetic materials impact society and the environment by...

 COLLECT EVIDENCE as you work through the lesson. Then return to these pages to record your evidence.

EVIDENCE

A. What evidence have you discovered to explain why climate, natural resource availability, and the economic conditions of a region limit the production and uses of synthetic materials, such as the ones found in cell phones?

B. What evidence have you discovered to explain how synthetic materials, such as the ones found in cell phones, impact individuals and society?

MORE EVIDENCE

C. What evidence have you discovered to explain how synthetic materials, such as the ones found in cell phones, impact the environment?

When you are finished with the lesson, review your evidence. If necessary, based on the evidence, revise your claim.

REVISED CLAIM

Synthetic materials impact society and the environment by...

Finally, explain your reasoning for how and why your evidence supports your claim.

REASONING

The evidence I collected supports my claim because...

What limits the production and use of synthetic materials?

Cell phones are made up of different synthetic materials. Can you find every synthetic material in every country around the world? Not every country has the same natural resources. In the next investigation, you will model the process of a country trying to produce a synthetic material from the resources found in your country.

⚙ENGINEERING INVESTIGATION

All Around the World

The world is in a race to build the tallest tower. Some countries have decided to submit a model of their design to the submissions committee. One of the criteria for each tower model is that it must be able to hold a marshmallow on top. First, gather the natural resources that your country has available. Next, build your tower. You will have 20 minutes. When the time is up the marshmallow must be on top of the tower. Your tower will be judged on:

- if it can hold the marshmallow, and
- how tall the tower is.

Country	Natural resources available						
	Binder clip	Wooden craft stick	Paper towel	Plastic bottle	Pieces of clay	Rubber band	Teacher advice
Russia	5	10	2	0	2	10	0
Mexico	0	7	2	0	2	8	0
China	8	0	2	1	2	10	0
United States	0	10	0	0	2	8	2
Brazil	5	8	0	0	0	8	0
Australia	5	6	0	0	0	0	0
India	5	2	4	0	5	0	2
Argentina	4	4	2	0	4	8	0
South Africa	8	0	0	0	2	0	2
Libya	0	0	0	0	5	10	0

1. Summarize your experience building your tower. What worked well, and what caused you problems? Define at least one design problem that your team solved. Evaluate your solution to the problem.

_____

_____

_____

_____

_____

_____

_____

2. How were the technologies that each country built similar, and how were they different?

_____

_____

_____

_____

3. What might have affected the outcome of the tower build?

_____

_____

_____

_____

_____

_____

_____

_____

EARTH SCIENCE ▸ **Connection** Humans depend on Earth's land, oceans, atmosphere, and biosphere for different natural resources. Many of the resources, including minerals, water, and biosphere resources, are limited. In addition, many of these resources are not renewable or replaceable over human lifetimes.

Natural Resource Availability Natural resources are distributed unevenly around Earth. For example, iron ore is found near the surface of Earth and is mined easily in some regions in China. Large amounts of iron ore are used in the production of steel, a synthetic material. In other countries, iron ore is rare and steel must be imported, or brought into, the country. Climate also plays a role in natural resource availability. One of Russia's resources is timber, because its climate is favorable for growing large forests.

Synthetic Material Production In Russia, it is easy to make synthetic materials from timber because it is readily available. The production and use of technology, such as synthetic materials, varies from region to region due to differences in climate and natural resource availability.

Economic conditions of a region also determine the production and uses of synthetic materials. During favorable economic conditions, less effort is spent on synthetic materials that provide for basic living needs, because those needs are taken care of. More effort can be put into synthetic materials that improve a specific aspect of living, such as entertainment.

THREE-DIMENSIONAL THINKING
Compare and contrast what you just read with your results from the *All Around the World* investigation. **Explain** why your tower was or was not a success. Record your response in your Science Notebook.

 Want more information?
Go online to read more about the impacts of synthetic materials on society and the environment.

COLLECT EVIDENCE

Cell phones are made where most of the raw materials needed to make them are found. How do climate, natural resource availability, and economic conditions limit the production of synthetic materials, such as cell phones? Record your evidence (A) in the chart at the beginning of the lesson.

What are the impacts of synthetic materials on individuals and societies?

The need for new synthetic materials can come from the society of the region. Often technologies are developed based on the needs, desires, or values of a society as a whole. Other technologies can be based on individual needs or desires. For example, phones developed as a technology to fill a desire for people to communicate long distances instantaneously. Cell phones are used all around the world. How do cell phones impact individual people? How do they impact societies?

INVESTIGATION

In the World

▶ **GO ONLINE** Watch the video *Light-Setting Casts* to learn more about how synthetic materials impact individuals and society.

1. Identify if the synthetic material filled a need or a desire. Explain.

_____

_____

_____

2. Identify if the synthetic material was for an individual or a society? Explain.

_____

_____

_____

3. What effect did the use of the synthetic material have on an individual or a society?

_____

_____

_____

Individual and Societal Impacts Synthetic materials have various impacts on individuals or societies depending on the function of the synthetic material. Ethanol, a renewable fuel made from plant material, such as corn, has different impacts based on the level—individual or societal. Gasoline mixed with ethanol is less expensive than pure petroleum-based gasoline. It is more cost effective to fuel a vehicle with an ethanol blend than with petroleum-based gasoline, reducing the cost of fuel for individuals.

Ethanol has other impacts at the societal level. Because ethanol blends are cheaper, it reduces dependency on petroleum-based gasoline. When ethanol burns, fewer pollutants are released. This reduces the amount of smog and acid rain created by vehicles.

COLLECT EVIDENCE

How do synthetic materials, such as those used in cell phones, impact individuals and society? Record your evidence (B) in the chart at the beginning of the lesson.

FOLDABLES®
Go to the Foldables® library to make a Foldable® that will help you take notes while reading this lesson.

What are the impacts of synthetic materials on the environment?

EARTH SCIENCE ⟩ Connection   The natural resources used to make synthetic materials come from the environment. Natural resources are obtained by extraction from Earth or harvested or farmed from the land or water. As societal demand for a synthetic material increases, the demand for the natural resources used to make the synthetic material also increases.

Copyright © McGraw-Hill Education   Roy Kaltschmidt/Lawrence Berkeley National Lab/USDOE

INVESTIGATION

Obtaining Resources

Palm oil is found in many different synthetic products. You might have used a product that contains palm oil today. It can be found in lipstick, shampoo, detergent, packaged bread, and soap. Along with corn, it can also be used to make ethanol for biodiesel.

Due to the demand of renewable oil sources, some countries have increased the farming of palm oil dramatically in recent years. The plant that produces palm oil grows best in equatorial climates that receive a lot of rain. Examine the two photos on the next page. They were taken only a year apart. What do you notice?

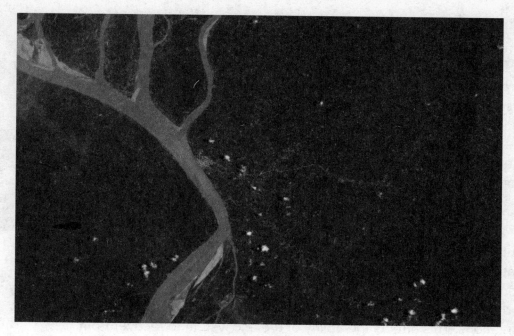

Tamishiyacu, Peru in 2012

Tamishiyacu, Peru in 2013

How did the rain forest outside of Tamishiyacu, Peru, change from 2012 to
2013? What effect might this have on the ecosystems in Tamishiyacu?

_____

_____

_____

_____

ENVIRONMENTAL Connection Sometimes obtaining a resource to make a synthetic material significantly alters the biosphere. For example, draining a wetland creates large amounts of land to be used for agriculture, such as growing corn. The corn is then used to make ethanol. Draining the wetland and planting a single crop—corn—damages the natural habitat of the organisms that lived in the wetland. This could lead to the extinction of some species. The long-term functioning and health of any ecosystem is influenced by its relationships with human societies.

The different methods that are used to extract, harvest, transport, and consume natural resources influence the many natural systems found on Earth. As you saw in the Investigation *Obtaining Resources,* the biological diversity and the composition of the land decreased when it was cleared for palm oil plantations. This affects the ability of the ecosystem to recover from effects of human activities. As consumption of the synthetic material increases, so do the negative impacts to Earth unless sustainable methods for obtaining the natural resource are developed.

Changes to the biodiversity of an area due to habitat destruction can influence other human resources. For instance, when rain forests are cleared to grow crops, the other food, energy, or medicine resources the rain forest provided are no longer available. Changes to an area affect how the biome cleans the water resources that the human populations drink. The changes can also affect how the biome recycles nutrients to maintain healthy soil for plant growth. As populations of organisms compete for limited resources, the growth and reproduction of organisms are reduced.

THREE-DIMENSIONAL THINKING

WRITING Connection A large deposit of bauxite, a mineral containing aluminum, was found in your community. The deposit crosses a nature preserve. A company would like to mine the bauxite in order to extract the aluminum for use in synthetic materials. **Develop an argument** on what could be some of the positive and negative impacts of mining the bauxite in your community.

Producing Synthetic Materials You have just learned how obtaining the resources needed to make synthetic materials impacts the environment. To make a cell phone, different synthetic materials must be made first. One material used to make cell phones is plastic. What are the impacts of making plastic? Let's find out!

LAB Making Plastic

Safety

Materials

milk

white vinegar

beakers (2)

graduated cylinders (2)

plastic spoon

coffee filter

funnel

water

Procedure

1. Read and complete a lab safety form.

2. Add 100 mL of warm milk to a beaker.

3. Add 10 mL of white vinegar to the milk. Stir for 1 minute. Record your observations.

4. Place a coffee filter in a funnel. Hold the funnel over another beaker and pour the milk-vinegar mixture into the filter-lined funnel. Observe the contents of the beaker and the funnel. Record your observations.

5. Remove any clumps from the funnel and rinse them in water. Form the synthetic plastic into a shape and place on a paper towel to dry and harden.

6. Follow your teacher's instructions for proper cleanup.

Analyze and Conclude

7. Identify the natural resources used to make the synthetic material in the lab.

8. Explain what had to happen for the synthetic material to form.

9. Did anything else form when you made the synthetic material? Explain.

10. How could you dispose of the products created in the lab?

By-Products In order for ethanol to be formed from plant materials, the plant materials have to undergo a chemical reaction. Sometimes when a chemical reaction occurs, a secondary product is also created. A **by-product** is a secondary product that results from a manufacturing process or chemical reaction. The chemical reaction that produces ethanol also produces carbon dioxide, a by-product.

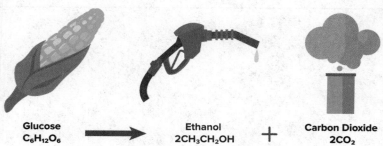

Glucose
$C_6H_{12}O_6$ → Ethanol
$2CH_3CH_2OH$ + Carbon Dioxide
$2CO_2$

A by-product can be useful and beneficial to the manufacturing process, or it can be considered waste. If a by-product is considered waste, it must be disposed. Disposal of waste by-products can have different impacts on society and the environment.

THREE-DIMENSIONAL THINKING

One way benzene (C_6H_6) can be made is by taking toluene ($C_6H_5CH_3$) and combining it with hydrogen gas (H_2). This chemical reaction is modeled by the following equation:

$$C_6H_5CH_3 + H_2 \rightarrow C_6H_6 + CH_4$$

Identify the by-product of the system. Explain how you know which substance is the by-product.

_____

_____

_____

You have learned that producing synthetic materials sometimes creates waste by-products. Those by-products must be disposed. But what happens when the synthetic material itself reaches its end-of-life and needs to be disposed? More than 125 million cell phones were thrown away in 2010. One natural resource used to make a cell phone is gold. That means 4,500 kg of gold is also thrown away with all those cell phones. Are there other ways to dispose of these resources?

Toss It

1. Examine each synthetic material supplied by your teacher. Predict if the material can be recycled, reused, or must be disposed of in a landfill or incinerator. Record your notes in the space below.

2. **WRITING Connection** Research to see how each material can be disposed of in your community. Update your findings.

3. Using technology, develop a way to inform your community about one method for reducing the amount of synthetic materials sent to a landfill or incinerator.

Disposing of Synthetic Materials Even though many organizations provide solutions for recycling or reusing cell phones, many still end up in a landfill or incinerator. Many synthetic materials are disposed of in the same way.

INVESTIGATION

Analyzing a Text

READING Connection You must be able to determine if information is credible, accurate, scientific, and unbiased. It is important that you can analyze a source for facts, misleading information, opinions, and describe how they are supported or not supported by evidence. This process helps you evaluate the strengths and weaknesses of information and make informed decisions.

CLOSE READING

Inspect

Read the passage *Single use, more like "multiple" use plastic bag*.

Find Evidence

Reread the passage. Underline two examples of a fact, highlight two examples of misleading information, and circle two examples of opinion.

Make Connections

Communicate With your partner, identify a claim made by the passage. Record two pieces of evidence that support the author's claim. What reasoning do they have for why the evidence supports the claim?

Single use, more like "multiple" use plastic bag

My state legislature wants to place a ban on plastic bags. The legislature claims it costs too much money to recycle plastic bags. The new law will require stores to charge people who use plastic bags, instead of using reusable bags.

The plastic bag is not just for hauling your purchased items home. I have found numerous uses for plastic bags. I always carry a plastic bag with my umbrella just in case it happens to rain. I can put my wet umbrella in the dry bag and not drip water everywhere. Last summer, I broke my arm. I covered my cast with a plastic bag every time I took a shower. I give my neighbor plastic bags for her litter box because I love her cat. The bags are extremely handy while I walk my dog, Zeus. Without plastic bags I don't know how I would pick up after my dog.

The group of people pushing the new legislation state the bags are harmful to marine life. For example, sea turtles confuse the plastic bags for jellyfish and ingest the bags. So only the bags that end up in the ocean cause harm. I'm not throwing my plastic bags in the ocean. My plastic bags end up in a landfill. I read polyethylene plastic bags contribute less CO_2 to greenhouse gases than paper bags and compostable plastic bags. I think this is more reason to continue using polyethylene plastic bags.

In conclusion, there are many possibilities for using plastic bags. I believe there should not be new legislation limiting companies for using plastic bags.

People have strong opinions on the production, use, and disposal of synthetic materials. It is important to make an informed decision by understanding all of the facts about how the synthetic material impacts society and the environment. In the next activity, you will evaluate sources of information to produce unbiased findings.

Impacting the Environment

1. **WRITING** **Connection** Choose one synthetic material. Research to obtain information on the natural resources used to make the synthetic material. Use at least two sources in your research.

In your research, include:

- methods used to extract or harvest the natural resources,
- methods used to transport the natural resources to a manufacturing plant, and
- how these methods influence the natural system surrounding the natural resource.

Be sure to evaluate each source for credibility, accuracy, and bias. Cite each source that you use. Record your research notes and your sources in the space below.

2. To communicate your findings, use a computer program to create a public service announcement (PSA) on how the production of the synthetic material impacts the environment. Edit and revise your PSA based on peer feedback.

Read a Scientific Text

Synthetic materials do not always have a negative impact on the environment. Some synthetic materials are developed specifically to minimize negative human impacts on the environment.

CLOSE READING

Inspect

Read the passage *Plants to Feed This—and Other—Worlds.*

Find Evidence

Reread the passage. Underline evidence of how the synthetic material helps the environment.

Make Connections

Communicate With your partner, discuss what the article gives as reasoning for how the synthetic material is beneficial to the environment. Would you invest in this synthetic material?

PRIMARY SOURCE

Plants to Feed This—and Other—Worlds

[...]

Growing plants in a spaceship, and one day on another planet, is a complicated endeavor. But one tool making it easier is a specially formulated fertilizer, developed years ago with NASA help, which has also drawn huge accolades from growers on Earth.

[...]

The fertilizer, created by Sarasota, Florida-based company Florikan, is coated in polymers that control when and how much of each ingredient is released over six months to a year. [...] The team's first major success, red romaine lettuce cultivated on the International Space Station in 2015, was the first produce ever grown and eaten in space.

Nutrients on Demand Florikan founder Ed Rosenthal did not expect to help grow vegetables in space when he first began developing his award-winning fertilizer. He just saw an opportunity to improve how nutrients are delivered to plants.

[...]

Traditional fertilizers are often applied monthly—requiring huge amounts of fertilizer and a large workforce to apply it. But Rosenthal knew that much of that fertilizer was never actually absorbed by the plant. As he told one grower friend, "I believe you're wasting more than two-thirds of your nitrogen: it's going straight into the groundwater." Florikan's new staged-release fertilizer would get the same results with a third of the fertilizer, and it only needed to be applied once.

[...]

A Boon for the Environment The key advantage to Florikan's staged nutrient-release fertilizer is that growers need to use far less of it, far less often than traditional formulations. That significantly reduces the harmful environmental impact of nutrient runoff, and it also means less labor and lower costs for growers. Nitrogen, in particular, has been linked to harmful algal blooms, which can release toxins that harm, and even kill, marine wildlife, including dolphins, manatees and sea turtles.

Source: National Aeronautics and Space Administration

⚙ ENGINEERING ▸ Connection Some synthetic materials have been designed specifically to help lessen the impacts that humans have on the environment. The development of new synthetic polymer-coated fertilizers is an example of the interdependence of science and engineering. The need to grow plants on the *International Space Station* started scientific research on time-release fertilizers. As the fertilizer was developed, it generated a new engineering field into how these fertilizers could be used in agricultural applications on Earth.

Engineering advances have led to important discoveries in virtually every field of science, and scientific discoveries have led to the development of entire industries and engineered systems.

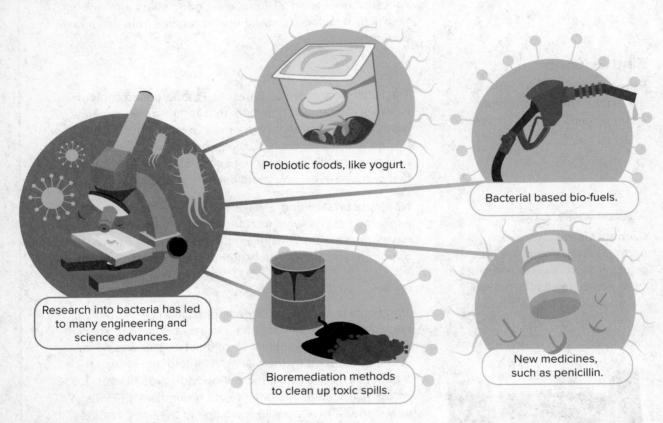

Probiotic foods, like yogurt.

Bacterial based bio-fuels.

Research into bacteria has led to many engineering and science advances.

Bioremediation methods to clean up toxic spills.

New medicines, such as penicillin.

Some people might wonder why synthetic materials haven't been developed to minimize every negative impact humans have on the environment. It is important to keep in mind that the development of new synthetic materials is limited by current scientific research. This research is driven by societal needs, desires, and values. Materials cannot be created from nothing. This is why we must rely on natural resources to make synthetic materials.

COLLECT EVIDENCE

How do synthetic materials, such as the ones used to make cell phones, impact the environment? Record your evidence (C) in the chart at the beginning of the lesson.

A Closer Look: Cell Phones

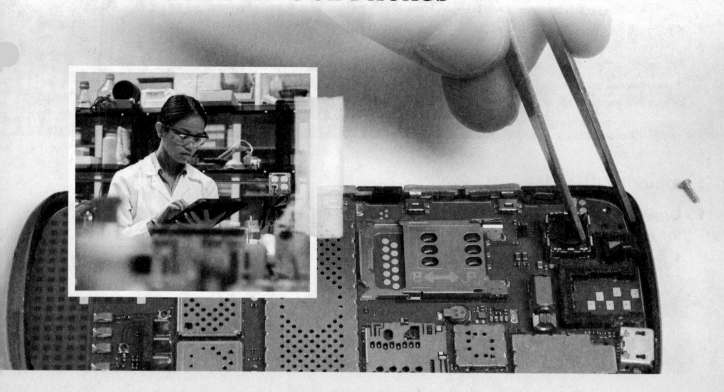

Cell phones are a tool that allows for instant communication around the globe. They can connect people from all walks of life from different countries. This allows humans to gain a better understanding of each other.

Cell phones can be used to help protect the environment. For instance, a cell phone allows a scientist collecting data in the field to quickly get in touch with his coworkers in a lab. This fast communication means if something changes, scientists can quickly contact each other to share information that might help protect the environment.

More than 125 million phones are discarded each year, contributing to waste in landfills and incinerators. Cell phones contain important natural resources that can be reused and recycled. Cell phones are approximately 40% metals, 40% plastics, and 10% ceramics and other trace materials. If you know someone who is planning to upgrade to a new phone, why not suggest they recycle their old one?

It's Your Turn

WRITING ⟩ Connection Many cell phones are used for an average of just 18 months before being replaced. Research ways your community recycles and reuses phones. Create a public service announcement informing your community about how they can dispose of their phones in a way that benefits society and the environment.

Summarize It!

1. **Organize** information about how synthetic materials impact society and the environment.

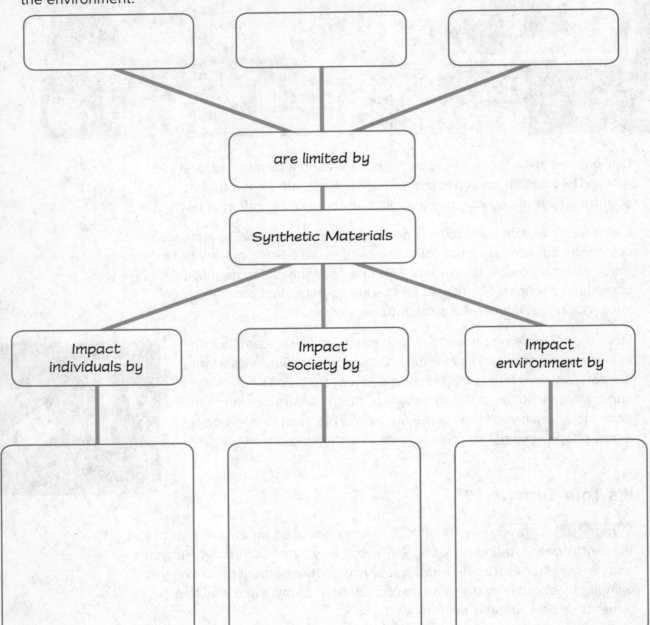

```
[              ]    [              ]    [              ]
        \              |              /
              [ are limited by ]

              [ Synthetic Materials ]
          /            |              \
[ Impact        ] [ Impact      ] [ Impact          ]
[ individuals by] [ society by  ] [ environment by  ]
     |                 |                 |
[              ]  [              ]  [              ]
```

Three-Dimensional Thinking

Read the passage. Then, use the passage to answer questions 2–3.

> Modern plastics are so durable that objects that are 50 years old have been discovered in the Pacific Ocean. These plastics come together to form an "island" of garbage. The garbage patch has been growing for 15 years and now covers an area twice the size of the continental United States.
>
> Some animals mistake plastic debris for food. Plastic debris causes the deaths of more than a million seabirds and more than 100,000 marine mammals each year. Plastic debris in the ocean also poses a risk to human health. Hundreds of millions of tiny plastic pellets—the raw materials for the plastics industry—are spilled every year, and work their way into the sea. These pellets attract many manufactured chemicals. The chemicals enter the food chain when animals consume the pellets of plastic. These animals are then caught and sold for human consumption.

2. What does the passage claim happens to plastics when they get to the ocean?

    A The plastic sinks to the bottom of the ocean.

    B The plastic floats along currents and ends up on islands.

    C The plastic is consumed by animals and enters the food chain.

    D The plastic is consumed by humans.

3. What could humans do to reduce the amount of plastic in the ocean?

    A Humans could reduce the amount of plastic that ends up in the trash by recycling.

    B Humans could reduce the amount of plastic that ends up in the trash by using plastic alternatives such as glass or paper.

    C Humans could reduce the amount of plastic that ends up in the trash by developing new technologies that filter plastic pellets out of the water system.

    D All of the above.

Real-World Connection

4. **Identify** a synthetic material and explain how your life would change if that synthetic material had never been developed.

_____

_____

_____

_____

5. **Explain** Not all synthetic materials are harmful for the environment. Explain how a synthetic material can positively impact the environment.

_____

_____

_____

_____

_____

 Still have questions?
Go online to check your understanding about the impacts of synthetic materials on society and the environment.

 **REVISIT SCIENCE PROBES**
Do you still agree with the student you chose at the beginning of the lesson? Return to the Science Probe at the beginning of the lesson. Explain why you agree or disagree with that student now.

EXPLAIN THE PHENOMENON

Revisit your claim about how synthetic materials impact society, the environment, and economies. Review the evidence you collected. Explain how your evidence supports your claim.

PLAN AND DESIGN
STEM Module Project
Engineering Challenge
Now that you've learned about how synthetic materials impact society, the environment, and economies, go to your Module Project to research technologies used to make shelters. Keep in mind what they are made out of and how they can be considered eco-friendly.

PERFORMANCE EXPECTATION

Take Cover

You work for a non-profit company that supplies shelters to people affected by natural disasters. Your company has asked you to be on a team that will develop a shelter that is economical and has a low impact on the environment.

You are tasked with researching what current technologies are used to make shelters and evaluating how well they met the criteria. You will then design your own shelter and test whether it can withstand a natural disaster.

Planning After Lesson 1

Before you begin designing your shelter, think about the following:

- who needs this solution
- what needs must be met
- any relevant scientific issues
- any societal or environmental impacts

Explain why it is important to define each of the above components.

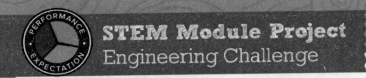

Planning After Lesson 1, continued

What criteria does this shelter need to meet?

What constraints limit your shelter?

Planning After Lesson 2

Research technologies that are used to make shelters. Record the questions and answers that drive your research in the space below. Use at least three sources. Include citations for your sources.

Create a table that lists the following:

- each type of technology that was researched,
- the materials that make up the technology, including the natural resources that the materials come from and the chemical processes used to form the materials from the natural resources,
- how the technology provides a shelter, and
- the societal and environmental impacts of each technology.

Add columns to the table that identify each criteria and constraint. Use the table to evaluate the strengths and weaknesses of each technology.

Using the results of the table above, make a claim about the effectiveness of each technology as a solution to the proposed problem. Support that claim with evidence and reasoning.

Create Your Design

Based on your research and evaluation of current technologies, design a shelter. Your shelter must withstand three of the four events:

- a wind storm
- an earthquake
- a heavy rainstorm
- extreme temperatures (hot or cold)

Sketch the design. In the sketch include the materials that will be used, the shape of the final product, and how this technology provides the best solution by addressing the criteria and constraints of the problem. Identify the three events the shelter will withstand and the features that will help the shelter withstand each event.

Once a sketch has been approved by your teacher, start building the model of your shelter technology as a group.

Test Your Design

Develop and use a systematic process to determine how well competing designs meet the criteria and constraints of the problem. Record your process and the results of your testing. On the basis of the testing, list one way you could modify your design to improve it.

Evaluate Your Design

Evaluate your solution against others that are designed to address the same events. Look at both short-term and long-term consequences, positive as well as negative, for the health of people and the natural environment.

What are some ways that recycled or reused materials can be included in your design?

Congratulations! You've completed the Engineering Challenge requirements.

Create a presentation to market your shelter. Use multimedia in your presentation. Share your presentation with your class.

Module Wrap-Up

REVISIT
THE PHENOMENON

Think about everything you have learned in the module about synthetic materials and their impacts on society and the environment. Construct an explanation for what the structure in the photo is made out of and if it is eco-friendly.

OPEN INQUIRY

What are one or two questions you still have about the phenomenon?

_____

_____

_____

_____

Choose the question that interests you the most. Plan and conduct an investigation to answer this question.

Glossary

Multilingual Glossary

A science multilingual glossary is available on the science website. The glossary includes the following languages.

Arabic
Bengali
Chinese
English
Haitian Creole

Hmong
Korean
Portuguese
Russian
Spanish

Tagalog
Urdu
Vietnamese

Cómo usar el glosario en español:
1. Busca el término en inglés que desees encontrar.
2. El término en español, junto con la definición, se encuentran en la columna de la derecha.

Pronunciation Key

Use the following key to help you sound out words in the glossary.

a	back (BAK)
ay	day (DAY)
ah	father (FAH thur)
ow	flower (FLOW ur)
ar	car (CAR)
E	less (LES)
ee	leaf (LEEF)
ih	trip (TRIHP)
i (i + com + e)	idea (i DEE uh)
oh	go (GOH)
aw	soft (SAWFT)
or	orbit (OR buht)
oy	coin (COYN)
oo	foot (FOOT)

Ew	food (FEWD)
yoo	pure (PYOOR)
yew	few (FYEW)
uh	comma (CAH muh)
u (+ con)	rub (RUB)
sh	shelf (SHELF)
ch	nature (NAY chur)
g	gift (GIHFT)
J	gem (JEM)
ing	sing (SING)
zh	vision (VIH zhun)
k	cake (KAYK)
s	seed, cent (SEED)
z	zone, raise (ZOHN)

English — B — Español

by-product/natural material

subproducto/material natural

by-product: a secondary product that results from a manufacturing process or chemical reaction.

subproducto: producto secundario que se obtiene de un proceso productivo o una reacción química.

G

groundwater overdraft: occurs when groundwater is withdrawn from aquifers faster than natural recharge can replace it.

exceso de extracción de agua subterránea: ocurre cuando el agua subterránea es extraída de acuíferos más rapido que la restauración natural

M

material: the matter from which a substance is or can be made.

mining: the process by which commercially valuable resources are removed from Earth.

material: sustancia que constituye la materia.

minería: proceso por el cual los recursos valorables son extraídos de la Tierra.

N

natural material: any physical matter that is obtained or made from plants, animals, or the ground.

material natural: materia física que se obtiene de las plantas, los animales o la tierra.

natural resource: part of the environment that supplies material useful or necessary for the survival of living things.

nonrenewable resource: a resource that is used faster than it can be replaced by natural processes.

recurso natural: parte del medio ambiente terrestre que proporcionan materiales útiles o necesarios para la supervivencia de los organismos vivos.

recurso no renovable: recurso que se usa más rápidamente de lo que se puede reemplazar mediante procesos naturales.

O

ore: a deposit of minerals that is large enough to be mined for a profit.

mena: depósito de minerales suficientemente grandes como para ser explotados con un beneficio.

P

permeability: the measure of the ability of water to flow through rock and sediment.

porosity: the measure of a rock's ability to hold water.

permeabilidad: medida de la capacidad del agua para fluir a través de la roca y el sedimento.

porosidad: medida de la capacidad de una roca para almacenar agua.

R

renewable resource: a resource that can be replenished by natural processes at least as quickly as it is used.

recurso renovable: recurso natural que se reabastece por procesos naturales al menos tan rápidamente como se usa.

S

soil: a mixture of weathered rock, rock fragments, decayed organic matter, water, and air.

subduction zone: the area where one plate slides under another plate.

synthetic material: a material obtained from a natural material that has undergone a chemical reaction.

suelo: mezcla de roca meteorizada, fragmentos de rocas, materia orgánica descompuesta, agua y aire.

zona de subducción: área donde una placa se desliza debajo de otra placa.

material sintético: material obtenido de un material natural que ha sufrido una reacción química.

T

technology: the practical use of scientific knowledge, especially for industrial or commercial use.

tecnología: uso práctico del conocimiento científico, especialmente para empleo industrial o comercial.

Glossary
Glosario

Italic numbers = illustration/photo
Bold numbers = vocabulary term
lab = indicates entry is used in a lab
inv = indicates entry is used in an investigation
smp = indicates entry is used in a STEM Module Project
enc = indicates entry is used in an Encounter the Phenomenon
sc = indicates entry is used in a STEM Career